CLIMATE
SOUL OF THE EARTH

ALSO FROM DENNIS KLOCEK AND LINDISFARNE BOOKS

The Seer's Handbook
A Guide to Higher Perception

CLIMATE
SOUL OF THE EARTH

DENNIS KLOCEK

LINDISFARNE BOOKS
2011

2011
LINDISFARNE BOOKS
An imprint of Anthroposophic Press/SteinerBooks
610 Main Street, Great Barrington, MA
www.steinerbooks.org

Cover photo copyright © by Fedorov Oleksiy (shutterstock.com)
Except where noted, images are by the author or in the public domain.
Cover and book design by William Jens Jensen

LIBRARY OF CONGRESS CATALOGING-IN-PUBLICATION DATA

Klocek, Dennis.

Climate : soul of the Earth/Dennis Klocek.

  p. cm.

ISBN 978-1-58420-094-9

1. Climatology—Statistical methods. 2. Time-series analysis. I. Title.

QC981.45.K46 2010

551.6—dc22

2010027825

Printed in China

CONTENTS

Satellite image of Earth's atmosphere
(NASA/JPL/UCSD/JSC)

INTRODUCTION

Tracing the etymology of the word *atmosphere,* one sees how, for the Greeks, there was a kind of commonality between the concepts of wind, breath, soul, air, vapor, and vital principle and the concepts of animating spirit, or soul.* Later, during the Middle Ages, the soul was known as "the air body" and considered the seat of consciousness of the one who possessed it. Likewise, the air, or atmosphere, was considered the soul body of the Earth, the living center of the influx of movements that arise when the other planets interact with Earth. History tells us that Pythagoras (570–c. 495 B.C.E.) was the first to "hear" and interpret the geometrical relationships of planetary interactions as celestial music. Subsequently, medieval philosophers believed that these influences, caused by movements of other planets and registering in the soul of the Earth, were manifestations of what they called "music of the spheres." Today, we call these harmonic impulses "climate patterns."

THE COSMIC MONOCHORD

The atmospheric impulses that arise through planetary motion are "musical" because they are mathematically rhythmic phenomena. They can be either harmonic or dissonant, depending on the geometry of the planetary relationships when seen as a musical interval on the great monochord of the cosmos. On the great monochord, philosophers considered the Earth's circumference and its 360° to be the fundamental Earth tone. They saw that a planet moving in a harmonic geometric ratio (3:4 or 90° of arc) in relation to another planet would instigate a harmonic response in the Earth's air body, or soul. Musically, the ratio of 3:4 produces the interval of the fourth above the fundamental. That fourth above sounds against the fundamental of 360° and gives the harmonic tone of the fourth above the fundamental Earth tone to the

---

* In Sanskrit, *atman,* meaning "soul," or "self," is derived from the Indo-European root *et-men* (to breathe or breath). Echoing this in ancient Greek, the word for "soul" is either *psyche,* from the Indo-European *bhes* (to blow), or *pneuma,* from *pnein* (to breathe), derived from Indo-European base *pneu* (to wheeze or to breathe). Breath may be seen as a kind of vapor, and in Greek the word for "vapor" is *atmos,* from which we get our word *atmosphere.*

atmosphere. Philosophers saw the quality of this interval as a kind of tension, or blockage. It arises when two planets stand in longitudes separated by 90° of arc. Today, if we were to assign such a quality to the climate at such a juncture, we would call this harmonic response an "enclosed high" or a "blocking ridge." There is a great deal of tension in a blocking ridge.

By contrast, philosophers considered a planet moving to a musically dissonant or unstable relationship (7:10 or 108° of longitude arc) to be a disturbing force on Earth's soul body. The ratio of 7:10 produces the interval of a tritone. When sounded against the fundamental of Earth's 360° circumference, the 108° angle creates a dissonant tone. In the language of today's climate studies, we would call such dissonance, or turbulent relationship, a "trough" or a "low-pressure area." When sufficiently strong or persistent, such angles of arc between two planets are often coincident with hurricanes, typhoons, and blizzards.

Using these kinds of ideas, the great astronomer Johannes Kepler (1571–1630) theorized that a harmonic disturbance of Earth's soul body results in weather disturbances, and that harmonic tension and blockages lead to clear skies. Kepler, in fact, wrote music for such planetary motions (*Harmonice mundi,* 1619), which takes its title from these ideas.

To bring this thinking into a contemporary setting, we must emphasize the difference between *weather* and *climate.* Those who study climate patterns see a great difference between weather and climate, depending on the size of the phenomenon and length of the period in which it unfolds. Weather events happen over periods of a few days and within spaces that may be as small as a parking lot or as large as a whole region of a country. Climate events, on the other hand, develop over periods of a week to centuries, and within spaces that may include several states or provinces to half a hemisphere. Climate study requires one's methods to be capable of maintaining coherence over very large scales of time.

Numerical modeling on computers has advantages for the short-term weather forecast, but such methods are much less reliable for longer-term climate forecasts. The purpose of this book is to demonstrate the possibility of another way of modeling that can aid climatologists, those who study climate rather than weather. Nonetheless, both weather and climate involve an underlying mystery related to the seemingly random, yet remarkably consistent, rhythmic events in the atmosphere.

The term *stochastic* describes the paradox of randomness leading to consistency. Often described as random, a stochastic system is often not random at all. Rather, it is composed of so many variables that pointing to one and saying that it determines something in the system is to deny the wholeness of all the events that contribute to the phenomenon. This describes a central problem in modern climate research.

THREE ELEMENTS OF A STOCHASTIC SYSTEM

Time Series

Three elements constitute a stochastic system. First, it consists of a "time series," which means the event has a sequence of random variables that define the phenomenon. Some variables have more probable influence than others have but may not be immediately observable. What we can observe is that the events occur in a recordable sequence. In climate research, the very essence of the work is to establish the precise timing of past events accurately, which allows some form of modeling for future events. This becomes an enormous task when the phenomena extend over a decade (the Dust Bowl) or even several decades (sea surface temperature fluctuations since 1976). Therefore, the first element of a stochastic system is the hidden coherence between a series of seemingly random time events.

Random Field

The second element in a stochastic system is a "random field," or "space domain." There are points in the space at which seemingly random phenomena are unfolding, but when closely analyzed reveal statistically significant relationships. The elements of a random field change continuously, making it impossible to say that one element causes another; yet random fields often exhibit what mathematicians call "strange attraction." This term describes the hidden attraction that various elements of the random field have for one another. This manifests when the points in stochastic fields move through space and time in mysterious though remarkably coherent ways, even when the connection is not apparent. It is up to researchers to discover the invisible forces acting on or within the random field to make it respond in particular periods. Researchers must do all of this without the help of deterministic, cause-and-effect processes of reasoning—the standard "this causes that" experiment. One cannot prove cause and effect in a stochastic system. There are too many variables interacting in non-discrete periods to be able to say that one thing causes another. This is, again, an accurate image of the central problem of climate research. The chaos of multiple trajectories of forces, all acting across long periods and vast random spaces, requires an approach different from traditional chemistry or physics. However, in all of this chaos, there must be something on which to build a model. This leads to the third element of a stochastic system, the element of type.

Element of Type: Strange Attractor Phenomena

To study a phenomenon such as commodity or stock market fluctuations requires sticking to a particular type of event. One may not be able to use an insight into a corn-market fluctuation to form a model for movements in the silver market or to a stock index. On the other hand, soybean-market patterns may provide some insight into other types of grain markets, even though certain links will likely remain incomplete. Therefore, to form a stochastic model for stocks or commodities, one must pay attention to uncovering harmonious type characters for the modeling procedure.

This does not mean, however, that we must use only grain markets to study grain markets. Climate anomalies sometimes drive grain markets, but only for a month or two each year. The rest of the stochastic fluctuations in grain prices result from the pressures of international trade, monetary unbalances, social unrest, and manipulation of those who need the products versus manipulation of the market by traders. These forces are all part of the stochastic background for soybean-pricing structures.

Such a factor can be used to study other runaway bull markets in which climate patterns contributed in a large way to the pricing trends. This includes a severe drought, such as the Midwest drought of 1988, when the climate aspects of the random field and time series became a strong force in understanding the fluctuations that formed a runaway bull market. In 1988, climate dominated all other types of factors, including a certain currency weakness and the political situation in China. That year, the climate type dominated the random field and time series. All points in the stochastic time series had a strange attraction to climate. Using this idea in a much broader way, climate is becoming a significant attractor type of element for many seemingly unrelated elements in many other fields.

In the West, climate research is emerging as an increasingly predominant strange attractor in the stochastic systems behind world events. This fact is one of the reasons for writing this book. However, we must ask: Why is climate the "soul of the Earth"? What does it mean to have a soul? These questions, while they include the concept of stochastic systems, must be amplified by a concept from another realm of science in which rhythm is seen to have a significant influence on physical and emotional processes. This is the realm of heart medicine.

Rhythm, Heart Medicine, and the Stochastic System

In the study of cardiovascular disease, the term *ischemia* refers to a lack of blood flowing into an organ. In a fatty heart, the clogged arteries leading to the heart muscle prevent the heart from getting enough blood to sustain its activity. Thus, a condition of ischemia sets in, and the result may be heart failure. However, surgeons and doctors who study this phenomenon also recognize "ischemic response," whereby the heart can be trained to tolerate ischemic events.

Before surgery to open clogged arteries, which may use small balloons inflated in the arteries to crush the plaque there, surgeons rhythmically stop and start the flow of blood to the heart. First, they deny the heart its usual supply of blood for a short time, and then they let it flow again. The next period of denial is a bit longer before the flow is reestablished, then they continue to extend the denial period. Doctors have found that, by extending the period (time series) of blood stoppages in the preoperative period, they can train the postoperative heart to tolerate a larger repertoire of ischemic rhythms, giving it a more flexible range of operating rhythms following the surgery. In experimental conditions, researchers have literally brought dogs back from the dead by training their hearts to tolerate extended ischemic events. The postoperative heart remembers something from the preoperative rhythmic protocol.

To say that the heart remembers its training may seem to some a bit anthropomorphic, but surgeons themselves talk of the heart remembering the intrusion of surgery and feeling resentful of such an aggressive invasion. Some doctors say that the heart must "get over" its memories, and that it does so in its own rhythmical time series. The question is: What influence drives the rhythmical time series of the heart? Is that influence a stochastic type or category of element that can be included as a variable in the research? What makes rhythm a stochastic type? In other words, is there an archetypal type of rhythm?

PLANETARY RHYTHMS AND THEIR INFLUENCE ON HUMAN BEINGS

I know a man who had quadruple bypass a few years back. He had experienced a great deal of suffering in his life and had refused or been unable to grieve. He prided himself on being strong and never shedding a tear, even when life brought him tragic circumstances. Such a profile provides a key to heart disease. As we will see, doctors know this as mental stress. Mental stress is not the result of some outside force. It is produced within the consciousness of the individual. We could say that it is self-induced by one's particular soul mood. In heart medicine, this has profound consequences for postoperative capacities.

Doctors had scheduled that man's quadruple bypass surgery for a few days before a total solar eclipse. As the date approached, the surgeon expressed a desire to attend a conference and wanted to change the date of the surgery to the day of the eclipse. To those who study such rhythms, such a shift of schedule would set off alarm bells. Eclipse days are profoundly significant in their time signature and often ominous. After much persuasion on my part, the man eventually refused the doctor's schedule change. The doctor canceled the surgery, went to the conference, and then returned a few days after the eclipse to do the surgery, which went well.

I had told the man that he should remain vigilant for twenty-nine days following the surgery. This marked the return of the Moon cycle to the point where it had been at the time of the surgery. The man told his surgeon of this idea, and the surgeon considered it absurd. However, as the Moon returned in its time series to the point where surgery had violated the heart, some mysterious, dare we say, stochastic process took hold of the heart and it went into arterial fibrillation. The man had to be rushed to the hospital. Not only did the violated heart act out this event on the anniversary of the monthly lunar cycle, but also, one year later on the Sun calendar, a remnant of indignant heart memory resurfaced with the same arterial fibrillation pattern and put the man at risk of another visit to the emergency room.

What would have happened had the surgery taken place on the day of the solar eclipse and the heart remembered that? This is interesting, because medical data such as a patient's EKG and blood pressure are stochastic domains. If you ask a physician what a standard blood pressure reading should be, they would probably offer a range of numbers. If your numbers are within the range, you would probably be considered "normal." Then, however, the physician might go on to qualify the numbers and describe situations and constitutions in which the range could be stretched one way or another and still be standard. Even when blood pressure is within the standard range, there may still be pulse complications that indicate a probability that the condition would modify the blood pressure numbers. As such, the art of medicine relies on probabilities instead of exact numbers for critical readings of the dynamics in the physical organism. Again, what has this to do with a soul? To continue the story of the man and his heart, after the first event one month after the surgery, the man began to cry every day when he awoke in the morning. He did not cry about anything in particular; he just cried. In the beginning, he sometimes cried several times a day. Then, with time, it was once a day until the second event a year later, when the wells of his sorrow seemed to dry up.

Much research shows that mental stress is a significant predictor of ischemia of the heart muscle, independent of diet, genetics, or exercise. Patients with mental stress-induced ischemia had longer and more frequent episodes of postoperative ischemic events that imperil the heart with random rhythms. Studies show that life stress, aggressiveness, and hostility are not associated statistically with prevalence of heart disease. However, self-induced mental stress is a clinical factor in heart disease. This type of stress originates and propagates in one's soul, independent of the environmental situation. We may ask: If the man had grieved over his tragedies in life, would his heart disease have progressed? In this story, only the man's sorrow could heal the soul of its self-induced stress. The heart, as a key organ in relation to the quality of one's soul life, assumes the burden of chaos in one's outlook on life.

HEART MEMORY OF SOUL EXPERIENCES

Let's return to the heart's memory as linked to the Moon. Why did arterial fibrillation occur when the Moon returned to the place it had been when the heart was opened surgically? Memory is the fundamental element in this relationship. All the elements of heart disease have to do with memory. The heart remembers the soul's experiences, and it does so rhythmically. It specializes in rhythms. In the story of the man's heart, it is interesting to note that his impaired health had revealed itself seven years after his estranged son had committed suicide. This would be a heavy burden for any soul, and it took seven years for this burden to work itself so deeply into the remembering that it could no longer tolerate those memories. The rhythm of seven years is fundamental in the human soul. We find it reflected in the names of days of the week, which arose from a perception of the planets' roles in daily life: Moon day; Tiu (the German god of war, or Mars) day; Woden (Mercury) day; Thor (Jupiter) day; Freida (Venus) day; Saturn day; and Sun day.

EARTH'S SOUL MEMORIES AS CLIMATE AND WEATHER PATTERNS

The idea behind this book is that Earth, as a living being, also remembers. Events remembered by the Earth form a time series. Returning to the image of Kepler and the Pythagorean monochord, the events that Earth remembers are the significant relationships that arise and dissolve with other planets in the solar system. They register in the Earth's soul as climate and weather patterns. The stochastic time series of the orbital relationships shape the response to the memory in the random field of forces that express Earth's soul mood. In this book, the random stochastic field of forces that expresses the memories of Earth is the climate. The stochastic type category that links the climate and the memories is the archetypal source of rhythm itself, the movement of the Sun, Moon, and other planets.

This type category is both a time series and a random field of space. We can effectively use this as an archetypal blueprint for studying the stochastic parameters of climate events. In this view, planetary movements represent biographical events in the Earth's life. Earth remembers them in the stochastic blending of time and space found in the planets' orbital periods and configurations. The case studies in this book illustrate how we can develop a different approach to climate study in which the modeling parameters are driven by motion-in-arc data from an ephemeris or star chart that is then woven into phenomenal events and sequences played out in the atmosphere. Through such modeling, we see the daily interplay between motion-in-arc increments and the constantly shifting patterns of the atmosphere in periods of time and spaces familiar to standard climatology.

From this perspective, we could ask: Is it so strange to include within the stochastic study of climate the parameters of the movements and time series in the field of the celestial spheres? As climate study becomes more pressing in human life, is it so strange to include these profound stochastic variables into the climate models? The original *stokhast* (Greek) was a diviner who saw the divine within the mundane. The tools of the diviner were a god-centered outlook. It involved a deep understanding of the geometry of the universe and a fundamental belief that Earth is a living and ensouled being whose soul is intelligent, who has a heart, who remembers, who dreams of evolving into something better, and who must suffer and grieve before it can heal.

Diviners of traditional Cabalism and Gnosticism see the Earth being as Adam and the crescent Moon as Eva, the rib of Adam. The planets are Adam's bodily organs: Jupiter is the liver; Venus is the kidney system; the Sun is the heart of the first human being, Adam. The Sun as the heart is the center of a confluence of forces constantly exchanged among Earth and the other planets. The forces of the planets are an outpouring of possibilities that create a random field of life support that has behind it a deep harmonic mystery. Of special importance is the effluent exchanged between the Sun and Earth, an outflow that is the source of life. To the diviner, an eclipse is the equivalent of a heart attack or an ischemic loss of the life-giving Sun effluent on the Earth. The Earth remembers such events and, in the post-attack period between eclipses, uses the memory of the eclipses as placeholders for unfolding archetypal climatic events.

The following chapters illustrate this concept from numerous perspectives, using thirty years of actual case studies. However, there is still one other important aspect of climate study.

THE DESTINY OF EARTH

This book considers what we call "climate," ultimately, to be an expression of the fundamental task of the Earth being. It expresses the relationship between the more inanimate and mechanized forces of the planet as a physical being and the more inward biographically evolutionary journey of Earth as an ensouled being embedded within the larger cosmic drama. The Earth is much more than rocks, water, gas, and heat. It is more than the combined forces in all of the living bodies it selflessly supports. The destiny of Earth as a being of cosmic import is woven intimately into the destiny of humankind. The destiny of the being of the Earth is the model for the intimate joys and fears that form the inner aspects of what it means to be human and to have a soul capable of evolving to higher states of consciousness.

Like human beings, Earth is a living organism that does not exist in a void. Its life is filled with interactions with other beings. For human beings, interactions with other beings in life form the content of their souls. For the Earth, interactions with other beings come in the form of significant geometrical events. Earth's interactions with other planets constitute the soul of the Earth. The context of Earth's soul life is the immediate environment of the Earth, filled with energetic planetary exchanges in constant fluctuation. It is the thesis of this work that these energetic Earth interactions are composed of relationships with other planets, with the other planets representing beings who are contemporaries of the ensouled Earth. This work shows how these exchanges are registered in the soul forces of Earth as weather and climate events. If we can see these links, planetary movements can provide support for long-range forecasting in a time when climate issues are becoming ever more important.

THE TASK OF THE HUMAN BEING IN EARTH'S DESTINY

At the center of this book is the idea that the climate crisis is a co-crisis shared by humans and the Earth as part of our mutual evolution into higher states of consciousness. Not only is Earth the source of our body, but the Earth also depends now on our efforts to shift our consciousness toward goals higher than self-satisfaction, entertainment, and consumption. Climate is the interface in which the results of our efforts to attain higher consciousness are displayed for all of the cosmos to see and evaluate.

In the past, people understood that the weather has a lot to do with the realm of the gods, with Zeus and his thunderbolts as a prime example. As the time of the gods receded in human consciousness, nature spirits living in the phenomena of wind and weather began to populate the natural realm for traditional and agrarian cultures. Nature spirits, or elemental beings, were linked to humans and the weather through the mutual interpenetration of the human body and weather systems with the forces and formative patterns of rotations of the elements. Magical practices for controlling the weather arose from those connections.

During the Renaissance, alchemists understood that nature spirits are manifestations of relationships between the stars and planets and the forces that animate human beings. If we wish to see this sort of relationship in action, it helps to observe plants in their rhythms of growth. During spring, the green wood shoots out of the buds when the Moon passes through the constellation of Taurus (an earth sign). People have observed this phenomenon many times over the years. When picking flowers and herbs or starting a fermentation process, it is good to notice the time when the Moon transitions from a water constellation into a fire sign. Fermentation is more successful when begun at that time of the month. Rose petals, when picked at that time, have a remarkable fragrance. Dew is heaviest during the month when the Moon passes through a water constellation. These relationships show links between the forces in plants and the motion of the planets.

The work of Li Guo Qing at the Institute of Atmospheric Physics in the Chinese Academy of Sciences in Beijing has shown a distinct fluctuation in the flow patterns in the Northern Hemisphere and the cycles of the Moon in its motion in declination. He tracked the orbital periods of the Earth on the atomic clock and found that the perturbation of the atmosphere by the motion of the Moon in its monthly cycle coincided with mass motions of the polar jet stream in rhythmic cadences that were predictable.

These examples infer that the lunar motions greatly influence both the air and the water on Earth, from hemispheric scales down to teacup scales. Because air and water make up a large part of our physical bodies, it seems reasonable to infer that planetary movements influence us, as well.

Ask any bartender, emergency room physician, asylum orderly, or police officer whether they think the full Moon has any connection to human behavior. Statistics show that the full Moon marks a period of intense human inner activity. Simply keeping a dream journal for a year will convince anyone that our inner life is connected to the periods of the Moon.

Going more deeply into this realm, we can do an experiment with startling implications. Take a six-foot length of PVC pipe that is three inches in diameter. We will use it as a kind of spotting scope. Choose a day when there are small cumulus clouds in the sky and a gentle wind. Pick out a smaller cloud in the center of a group of clouds. Supporting your spotter scope on a fence or tree limb, aim and look through the tube at the chosen cloud, making it the center of the field of vision. It helps to imagine a connection between yourself and the cloud and that the cloud is streaming energy toward you as you observe it. If you can keep your concentration on the cloud for a few minutes, you will notice that the cloud you have chosen starts to disappear from the group, while the others stay intact. This is not an illusion and anyone can do it.

Concepts such as these were not wasted on the people of the past. Weather magicians, rainmakers, and others understood that an intimate link exists between the human elements and soul life and the elemental beings that animate natural phenomena and the motion and rhythms of the planets. Those diviners knew that the world is whole and that earthly phenomena are populated by beings who, like humans, originate in the starry realms and retain links to the planetary spheres by virtue of having physical bodies. Such understanding of the relationship between the Earth's soul and forces and human souls and forces were lost in the harsh light of empirical science. The consciousness needed to operate in stochastic systems such as medicine and climate study points in a direction different from that of intellectually based empirical research. Our scientific culture, driven by the protocols of computer modeling, finds that the divining consciousness needed for human beings to make a clear weather prediction requires the creation of an ensemble of many different numerical computer models. Out of the fluid fluctuations and biases of the many models converging and diverging on a solution, forecasters must still introduce intuition and an artistic, stochastic-probability approach to data organization in order to gain insight. In the end, a successful forecast is akin to a divining.

The destiny of the Earth depends on human beings becoming diviners again—that is, again seeing the divine at work on the Earth. One way to experience this is through the stochastic study of how planetary movements other than the Moon impact the time series of Earth and alter the random field of the atmosphere into floods, droughts, hurricanes, or benign and harmonious forces.

To see the Earth in this way requires dissolving the present fundamentalist hubris and immature inflation that drives world events. It requires the realization that there are forces far beyond our understanding that lurk as unknowns in natural systems. These unknown forces are intimate with us as beings with souls. The door is open to our understanding of these mysterious forces if we include the stochastic patterns present in the complex rhythms of the planetary movements.

In the past, the Earth soul moved us along our collective path of development. At that time, the divine was experienced in the everyday. North American First Nations peoples call this "living in a sacred manner." The challenge is still to see the divine at work in the life of the Earth. What has changed is that Earth's destiny no longer depends on a collective cultural worldview. The new possibility for humanity is that we each have the opportunity to make this shift in consciousness out of autonomous insights.

Planetary realms, with their intricate and remarkably consistent movements, can serve as a rigorous blueprint for this monumental shift in consciousness. We find a mirror of these divine rhythms in the great mystery language formed by the movements of planets, which manifest as climate, the soul of the Earth. It is this author's belief that the present climate crisis is a call for human beings to divine once again the spiritual in the manifest world. A great link between the cosmic dimensions of this looming global challenge and the patterns that manifest as weather and climate is the backdrop to planetary rhythms. If climate is the soul of the Earth, then the spirit of the Earth lives among the stars and planets.

Cumulus rain cloud (by Fedorov Oleksiy, shutterstock.com)

CHAPTER I

CLIMATE PATTERNS

THE PHYSICS OF CLOUD FORMATION: A WEATHER PRIMER

The most fundamental force pattern in cloud formation is the convection cycle. The first image shows this as a governing dynamic in all clouds. A convection cycle has two components: an expanding and lifting current of warm air moving in the center and contracting, and falling currents of cool air descending from

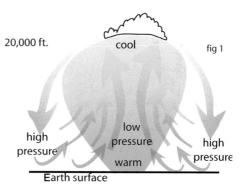

20,000 ft. cool fig 1

high pressure low pressure high pressure

warm

Earth surface

aloft along the outside of the cloud formation. Within the warm air rising in the center, the pressure is lower than within the cooler descending air on the outside of the convection cell. This occurs because, as the air rises, there is less of it to fill the space. Less air in the space means less air pressure. As the warm air rises, it expands and cools. By the time it has arrived at the top of the convection cell it is very cool and begins to sink. As it sinks and fills the space below, the pressure rises because there is more air filling the space. More air means more air pressure in the space below the currents of descending cool air. In the final stage, the descending cool air, which is increasing the air pressure in the space, begins to warm as the pressure increases. The warm air then starts to rise, and the cycle begins all over again. Clouds are born at the top of the rise. An image of a current of air moving up a slope illustrates this pattern.

Air at the surface of the Earth is under greater pressure than air is at a higher altitude. When a current of air that contains no water vapor moves up a slope, it moves into an area of lower pressure. This causes the rising air

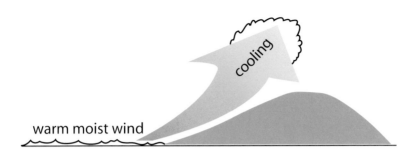

current to expand, since it is now in an environment of less pressure. All rising air expands into its environment. As we saw in the gas law, the physics of this expansion creates temperature differentials. Initially, the rising air cools as it rises and expands. While cooling, it borrows warmth from its environment and, in the process, becomes warm again and the cycle continues.

CLOUD FORMATION

Using the ideas in cloud physics, we can now look at the formative principles in different cloud types. In this image of a cumulus cloud, we see currents of warmth rising through the middle of the large cloud mass. The cloud itself is the water vapor discharged by the relationships between the gasses and the water. At the top of the cloud, there are invisible currents of cold coming down the outside of the cloud form. The rising and descending currents of warm and cold are part of a "convection cycle," or "convection cell."

All clouds have the convection cycle as the basis for their formation. A cloud is composed of droplets that have condensed from vapor as the air borrows warmth to balance its expansive cooling caused by reduced pressure. The form of the cloud reveals the forces that formed it in the many variations of the convection cycle. The following images illustrate the fundamentals of cloud formation.

Stratus clouds before dawn

We see a bank of flat clouds against a dark horizon before sunrise. These are called "stratus clouds" because they are arranged in strata, or layers. They show that the air close to the ground is cool before dawn. The moisture is not moving vertically in a convection cell, but sliding along the cool ground. In the lower left corner, we see some ground fog still lying in the valley. The little bank of wispy clouds above the stratus bank shows some effects of the moisture rising off the valley floor as it begins to form the first tendency toward vertical growth that characterizes a convection cycle.

Stratus clouds at dawn

The next image, a few minutes later, shows the stratus layers beginning to expand upward. The wispy band shows a telltale columnar pattern of convection. The little cloudlets within the cloud are rapidly expanding and lengthening upward to form columns. This pattern of growth will be greatly enhanced once the sun has risen and provides warmth to the clouds for their convection.

Opposite top: Here comes the Sun! The wispy bands are now much more substantial. The stratus bands against the horizon are massing up and forming a bank of clouds. Above the growing wispy band, another finer layer of very delicate gauze-like clouds is coming into form, like curdled milk congealing on a glass slab.

Opposite page, bottom: The middle layer of wispy clouds is disappearing as the fog on the valley floor rises. Both of these phenomena point to increased vertical motion. The "curdled-milk" clouds in the upper part of the picture are now gaining substantiality, even as the lower clouds are disappearing. Often, in the atmosphere, a bottom layer's warmth and moisture is sublimated upward to enable the unfolding of the upper layer's formation. In

Clouds begin to grow vertically (top); lower clouds sublimating (bottom)

Stratocumulus formation (top); altostratus formation (bottom)

Cumulus cloud banks forming

other words, the lower clouds disappear even as the higher clouds are forming out of their sublimated water vapor and warmth.

Opposite top: A few minutes later the upper layer has stabilized and the middle layer once again grows as the Sun begins to warm the atmosphere. The middle layer is now being fed water vapor again from the ground as the action of the Sun becomes more pronounced. The fog in the lower left is spreading out along the creek in the valley, feeding the cloud formation as the warming air draws up the moisture that had settled overnight.

Opposite bottom: The Sun has risen. The fog is lifting to the treetops. Currents of warmth carry moisture into the middle layer as the upper layer now sublimates. Warmth and cold are working into each other to balance the forces in the atmosphere as the middle layer grows substantially.

The Sun is now fully risen and warming the atmosphere vigorously (image above). The wispy clouds have risen much higher now, and the middle layer is drawing up the rising moisture from the Earth and forming it into dense masses. The stratus form of the earlier sequences is no longer as visible, but

Cumulus clouds beginning to drift

the cumulus, or heaped, cloud pattern is dominant. The fog has spread above
the trees and is feeding this rapid growth in the sky above

Cumulus clouds have massed and are now drifting on the wind. High
above, sublimated moisture is forming ice clouds while the fog completely
obscures the valley floor as it lifts heavenward to form new clouds.

Weather Vanes

For centuries, farmers and gardeners have looked to the heavens for accurate
indications of upcoming weather patterns. Storms are fluid and appear to be
chaotic phenomena. To predict them requires a strong memory of past events
linked to a clear weather eye for present conditions. Traditional weather
prophets also relied on nuggets of weather wisdom passed on from generation
to generation. These sayings are often just gleaning from an accumulation
of local and regional observations. Other sayings carry very astute insights
related to larger patterns and climatic sequences. Still others are sheer nonsense, such as "When the hen crows it will rain."

Behind the best and most reliable rhymes is a keen observation of what
could be called the rotation of the elements or, as the alchemists would say,

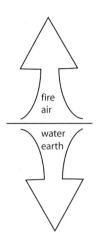

Stable elements

"the fire under the earth." This alchemical phrase refers to an unstable relationship among the four elements: earth, water, air, and fire.

In a stable condition, the earth and the water elements are interconnected through the force of gravity. Water moves down to meet the earth. The air and fire elements are interconnected through the influence of levity; warm air rises. Earth and water move down, while air and fire move up. In this stable form, there is a universal relationship. Water sitting in a puddle or in a pond is an image of this pattern. Earth is found below water. It is the element influenced most by gravity. Water sits on the earth. This is the condition of the atmosphere in the night. It is stable.

Sitting on the earth, water has a relationship to gravity, but it can also escape gravity. This occurs at the surface of water, where the elements split. Under the influence of warmth, water leaves the earth condition and joins air and fire as it moves upward. This is the condition of the atmosphere at night. In the morning, dew is present as a gift from the heavens.

To alchemists, the formation of dew was a miraculous and potent event. This image shows people gathering dew to make alchemical medicines. However, dew is also a weather indicator. There is an old rhyme that states, "Morning dew means weather fair / When dew is gone, a change beware." This accurate weather vane describes the stable conditions around the nighttime. In the evening, the day's warmth departs from the earth and rises to a higher level in the atmosphere, carrying moisture upward with it. As the evening progresses, the earth cools the air above and the cool air settles onto the

Gathering dew

earth. The water vapor that was carried aloft the previous day settles out in the coolest part of the night, the time just before sunrise as the Earth breathes in. The dew falls onto the earth, and the cycle starts again. The settling out of the earth and water below with the warmth above is the most stable pattern in the atmosphere. When this stable pattern dominates the evening, the

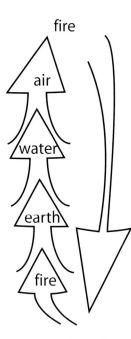

Fire under earth

weather tends to be calm and stable. As a result, the old rhyme about morning dew is a clear and accurate predictor of fair weather.

When the elements are arranged with the earth below, the water above it, the air above water, and with the warmth on top, fire is above the earth and the pattern is stable. With the approach of a storm, the air pressure lowers as warmth rises from the Earth and the Earth breathes out. This is disturbing to the stable pattern. Water is pushed up into the higher levels of the atmosphere. This allows the vapor in the air to remain in a state of levity during the night instead of settling through a calm atmosphere. In this state, water vapor is rising up to form clouds as a storm approaches. As a result, when the dew remains up in the air and does not fall, this is often a sign that bad weather is approaching.

When the water vapor stays up in the atmosphere, this is the opposite of a stable condition. When the atmosphere is unstable, this condition is known alchemically as "the fire under the earth." This situation is depicted in the third diagram. In this pattern, the fire, which is usually highest, is below the earth and all of the elements are driven upward. Alchemically, this condition normally arises during daytime, usually around 4 p.m., when the sun has warmed the earth and water enough that thunderstorms can begin. The stable relationship between the elements is disrupted, and the water in the earth begins to rise into the air as water vapor. The water vapor begins to evaporate into gasses as gases seek the periphery. Each element is driven upward by the activity of the fire below the earth. This unstable motion is the central phenomenon of storm activity. It is a result of a displacement of the archetypal condition of the elements.

An old weather rhyme that describes this pattern is "Thunder in the morning / All day storming." This describes a condition in which the earth in the morning is already warm, and there is an upward motion into the atmosphere. The Earth usually starts to breathe out in the morning around 4 a.m. and

Stratus formation

reaches a peak at about 2 p.m., just in time to give rise to a standard thunderstorm at 4 p.m.

Sometimes, storm conditions develop overnight, usually through the influx of a cold air mass into an area where the ground or water is already warm. This places the fire under the earth much earlier in the day, making the morning exhalation of the Earth more extreme. At such times, thunderheads are already in place by 10 a.m., and thunder in the morning signals unstable upward motion in the atmosphere throughout the day. These two polarities of stable and unstable atmosphere are the basis of all meteorological phenomena.

The different cloud types that arise at various times in the sky are really combinations of these two polarities. The stable pattern of the elements produces the stratus type of cloud (above), and the unstable pattern produces the cumulus cloud (next page). In the stratus formation, a warm air mass glides over a cool surface. In this pattern, the warmth is above and the cold is below. The resulting cloud is flat and moves horizontally across the cold earth or

Cumulus clouds beginning to drift

water. The warm air in the cloud is cooled by its passage across the colder layers below. As a result, earth and water remain below, while air and fire remain above. Precipitation from a stratus cloud is drizzly, and these types of conditions are usually accompanied by fog. This condition is exemplified by the old rhyme: When clouds descend / Good weather ends." The stratus cloud comes down to earth from above. In fact, fog is a cloud on the ground.

The opposite condition occurs when heaps of clouds pile up, one on top of the other. This is an image in which the warmth below drives the elements up into each other. The rapid upward motion leads to extreme winds and rain within the thunderhead. Hailstones driven upward by the rising warm air melt on the surface and collect rain. The hailstones are next driven up into the very cold air at the top of the cloud, where they freeze. The hailstones then become heavy and fall back through the rain bands, where the drops freeze to its surface until it is again within the lower regions of the thunderhead. This cycle may happen many times, until a hailstone is the size of a baseball. The rains that fall from these dynamic clouds are usually ice crystals that have melted in the warm updrafts and then fall out of the leading edge of the large cloud.

Stratus sunrise

This is why they are such large raindrops. This is also why flash flooding is a concern when the fire is under the earth.

The polarities of stable and unstable elements in the atmosphere are played out in other ways besides the two archetypal cloud forms. In the middle layers of the atmosphere, similar elemental relationships exist as a daily drama even when the weather is settled. Early in the morning, it often happens that the dominant cloud pattern during sunrise is composed of stratus clouds. This is because the dew has fallen and, even though the Earth is breathing out, that outbreath at dawn is only rarely capable of building cumulus clouds. At sunrise, the predominant cloud pattern has warm upper air descending down to a cool earth and water. If the ground is very cool, then the cloud may actually be on the ground accompanied by light drizzle. In such a case, this old rhyme holds true: "Rain before seven / Dry by eleven."

By ten, the stratus clouds usually show signs of losing their smooth outlines and begin to appear rough and fluffy on the edges. This is a sign that the fire is under the earth within the cloud itself. The cloud starts to absorb warmth, and vertical unstable currents begin to alter the shape. Usually, by

Stratocumulus (above); thunderhead (below)

noon there is a distinct cloud form in which the flat stratus shape grows upward. This cloud, a stratocumulus (opposite top), shows that the sun is warming the water in the stratus cloud and that vertical forces of levity are taking over from the gravity forces that dominated the stratus cloud at sunrise.

In warm climates near an ocean, the stratocumulus clouds can be seen throughout night, since the water below is so warm. Regular thundershowers in the afternoon are a standard feature in warmer climates. Warm updrafts carry water vapor from below to great heights. The cumulus clouds build on each other, and a tremendous rush of warm air pushes an anvil-shaped towering cloud into the upper atmosphere, where the vapor turns into ice (opposite bottom). The cool air above then sinks and becomes pressurized. As it pressurizes, it warms and starts to rise again, feeding the thunderhead with more warmth. This picture shows a remarkable up-rush of air and the spreading of the ice anvil at the top of the cloud. Hail and severe winds may accompany such a cloud. All of these modifications happen in the middle layer of the atmosphere. There are, however, modifications of clouds in the highest layer of the atmosphere that also follow these patterns.

The highest type of cloud occurs in the upper reaches of the atmosphere, where the cold and arid environment puts a lid on the weather. This area is known to meteorologists as the "tropopause," the place where the weather pauses. The most typical cloud to form in this layer is the cirrostratus. It is a stratus layer at the very top of the atmosphere. It resembles a fine blanket or veil of ice crystals. In this layer, any water vapor is transformed into ice crystals in the form of fine needles, which are electrically charged with one end positive and the other negative. In alchemy, the negative charge giver is salt, or in the old language, "glass"; the positive charger is sulfur, or "resin" in the old language. Opposite charges attract each other, and similar charges repel. In the high ice cloud, this means that the sky is covered with untold numbers of minute ice crystals, each locked in a push/pull relationship with the one next to it.

The entire sky may be filled with such a charged veil of ice needles, all lined up parallel to each other because of their charges. Picture an area hundreds of square miles filled with minute prismatic crystals, all facing in one direction. Now place the Sun or Moon behind this ice fog and the whole sky becomes like a Fresnel lens. This is the source of many rhymes regarding the

Cirrostratus bank (top); ring around the Sun

Feather cirrus

ring around the moon: "A Moon with a circle brings water in her beak." Or, "A ringed Moon means rain in a day or two." This saying becomes very accurate it we add, "When the wind is shifting and the barometer is falling."

These high ice clouds are known as cirrus clouds, which in Latin means "feather." The tendency in the cirrus cloud is for the needles to repel each other as the charge in the cloud becomes stronger. As the cloud grows from one end and spreads out in space, the needles increasingly repel each other and the cloud separates into fine streams, or feathers. The streams of ice crystals are sensitive to upper winds moving in front of storms. As a result, the cirrus clouds are prophets of far-off weather. They are usually formed along the leading edge of an incoming storm, like whitecaps on the tips of wind-tossed waves. They are runic clouds, often looking like hen scratches, mares' tails, or ancient glyphs in the sky. These clouds tell of strong winds high up . If the ice clouds form a ring around the moon or look like spilled buttermilk, it shows that the winds aloft are not violent.

As for rings around the Moon, the old sayings are "When the ring is far, the storm is near," or "The bigger the ring, the farther the wet." These sayings refer to the height of the cirrus layer. If the layer is very high up, then the circle of the ring will be smaller, since the cloud bank is closer to the source of light. If the layer is lower, in what is known as a "lowering sky," then the ring will be larger and, thus, the storm will be nearer. If later in the day the high ice-stratus veils begin to evolve into streaks or ripples, then these cirrostratus clouds are saying that a storm front is approaching the observer.

The lowering sky is the old term used to describe the approach of a storm seen as an evolving sequence of ice clouds. It usually begins with a cirrostratus ring around the moon, and then the flat veil of the cirrostratus evolves into a buttermilk sky. The soft clouds of the buttermilk sky (opposite top) are saying that there is a warm front approaching, and that the sky will most likely begin to lower. These are created when large masses of warm air float gently upward to a great height, causing the cirrostratus veil to form cauliflower masses of gently rising air in a cellular form. In the buttermilk sky the cirrostratus veils are influenced by rising warm air to create cells of curdled clouds. The soft clouds are actually very high ice clouds, undisturbed by horizontal winds. They are a very early harbinger of a change in the weather.

The next clouds in the sequence after the buttermilk sky are usually the formations known collectively as mackerel sky (opposite bottom). The old rhyme is "Mackerel sky, mackerel sky / Dry turns wet, and wet turns dry." The mackerel sky reveals that there is a high-altitude wind rippling the high ice layers. The mackerel sky looks like the stripes on a mackerel. It is made up of parallel bands of clouds that resemble ripples in the sky. They foretell the close approach of a storm system. In the cirrocumulus clouds, the earlier high ice sheets have become organized by a strong horizontal wind source aloft. This type of wind is most often being moved upward in the circulation around an approaching storm. When the old sailors and fishermen saw these clouds, they began to keep a careful eye on the direction of the wind. Mackerel sky is a lowering cloud formation that tells of a change in the prevailing weather patterns.

When the ice crystals in the mackerel sky encounter a warm upsurge from a distant low pressure area, the crystals that make up the cloud turn again into water vapor. As a result, the clouds become heavier and form lower to the ground. These lower clouds are composed of water vapor, unlike the cirrostratus, which are composed completely of ice crystals. One can tell the difference because the ice-cloud area always accompanied by optical or prismatic

Buttermilk sky (top); mackerel sky, or cirrocumulus (bottom)

Mouton, or altocumulus (top); nimbostratus (bottom)

effects, while the lower types, being composed of water, do not produce optical effects. This generally means the watery clouds are lower than the ice clouds. As the clouds lower, the next layer in this evolution from cirrocumulus or mackerel sky, is the "mouton," or sheep, clouds (opposite top). They resemble a flock of sheep grazing in a heavenly pastures in the middle atmosphere. In these clouds, the ice has turned to vapor, and the sky is lowering into rain.

The ultimate low cloud is the rain, or nimbus, cloud (opposite bottom). In this evolution, the nimbus is usually found in the form of a nimbostratus, or rainy stratus, cloud. These clouds bring drizzle and warm temperatures for a few days. They are part of what we know as a "warm-front sequence." We will look at fronts later.

Understanding Old Weather Rhymes
Today's Explanations of Yesteryear's Weather Rhymes.

As mentioned, many old weather rhymes contain great wisdom, while others are pure nonsense. How do we tell the difference? Doc Weather can provide some tips on how to look at the weather as described in old rhymes by describing the patterns behind them.

Looking at a weather map of North America, we can observe two fundamental patterns in the atmosphere. One is the high-pressure cell; the other is the low-pressure cell.

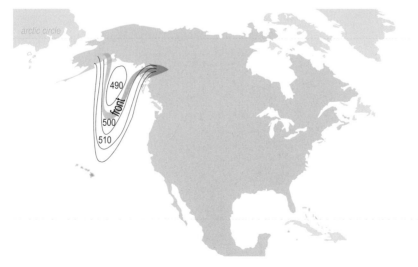

The first chart shows a low. Thin lines that reach down into the central Pacific just east of Hawaii define the general area of low pressure. The lines represent pressure lines, or isobars, that describe the various zones of pressure surrounding the low. The lowest pressure is in the center of the low (490mb), and the pressures gradually get greater moving out from this center (500 and 510mb). "Mb" stands for millibar, a unit of atmospheric pressure. When seen together, the different pressure zones around a low create the trough-like form in the atmosphere.

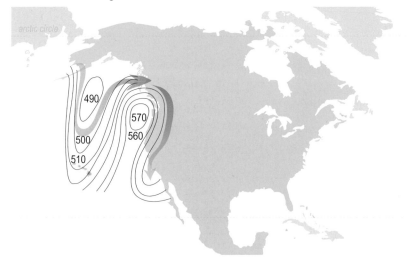

This trough is over the ocean and runs between the Aleutian Islands and Hawaii. Although the center of the low is in far northern latitudes, it is influencing the air pressures in areas far to the south. The wind circulation around the low exerts this influence. The chart adds the jet stream winds that guide storms across the hemisphere. The large arrow is the jet stream. The jet stream is a river of rapidly moving air in the upper layers of the atmosphere. It forms where high- and low-pressure areas interact with each other. It is a sensitive membrane between pressure cells. In the chart, the jet is moving around the low in the 500mb zone. Here, it would interact with a high-pressure area to form a border, or front. The winds around the low are cool on the western side and warm on the eastern side. This happens because they come down the western leg from the Polar regions and then move up on the eastern leg from tropical regions. Storms track the jet streams as they move across the Earth. Owing to the Earth's rotation, the circulation around a low is counterclockwise in the northern hemisphere. The chart shows this counterclockwise circulation around a low, depicting wind coming from the Bering Sea south to Hawaii and from Hawaii northeast into the Anchorage area.

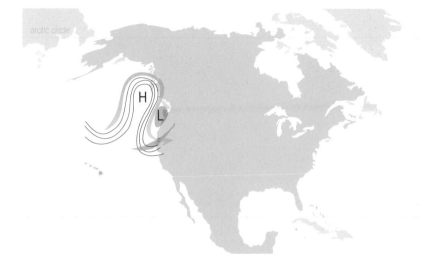

In this figure, a high-pressure area has been added to the chart in the eastern Gulf of Alaska. Meteorologists know this formation as a ridge. Because of the Earth's rotation, the winds around a ridge, or a high-pressure area, rotate in a clockwise direction in the northern hemisphere. In this chart, the ridge extends from the latitude of Hawaii north to the Canadian Rockies. This ridge prevents the winds circulating around the low in the western part of the Gulf of Alaska from entering the United States at a low latitude. This is because, even though the 570mb high in the ridge is centered over the coast of California, it is connected to another 570mb high over Hawaii. The ridge connecting the two is pushing the jet stream to the northeast. This kind of linking is called a "tele-connection." The clockwise circulation around both highs links up to create a mega-sized high with a strong current of westerly winds along the northern side of the linked highs (arrow). The jet stream around these types of linked highs is often very strong, owing to the long distances the winds travel around the highs. With a ridge as strong as this, the low in the Gulf of Alaska will not be able to flow south into California. The weather will be clear and sunny there, but there will probably be some weather in the Pacific Northwest as the cold air comes down from the north.

The last feature of this figure is a green wedge, known to sailors as the "dangerous semicircle," the place where the circulation around the low meets the resistance of the ridge. The winds here are stronger than any other place in the chart. The dangerous semicircle and the front line from the low are actually the same place, where the wind and rain are strongest as the two systems work out their energies. We will see the dangerous semicircle later. With these two patterns as examples, it is possible to make sense of many old weather rhymes about the wind.

The first such rhyme is very old: "When the wind is from the east, 'tis not fit out for man nor beast. But wind from the west makes weather the best." This describes the circulation around a linked high and a low. Figure 3 shows a low crossing the Gulf of Alaska on a steep pitch from the north to south. The low is a small, but it has a strong circulation. The winds in central California (black dot) are coming from the northeast (small black arrow) in L1, with the storm center off the coast of Washington. In L2, the wind shifts to the southwest (small black arrow) as the low approaches the coast, with the center off the coast of Northern California. In L3, the winds are from the east as the low turns eastward to cross the West Coast, with the center crossing the coast of central California. The storm winds will come from the east in Sacramento as the storm hits.

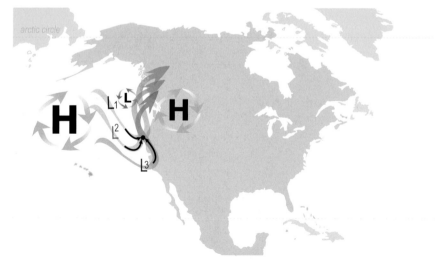

In this chart, cold air from a ridge in the western Gulf of Alaska is moving down to feed cold into the backside of the low. The front side of the low is digging into the California coast. As mentioned, sailors know this digging area as the dangerous semicircle. The chart shows that the jet stream winds are moving up into the western side of the continental ridge in western Canada and hitting British Columbia from the west. Weather there will be "the best," or clear and warmer than normal. On the other hand, the cold air from Alaska is coming down the eastern side of the ridge and funneling into the western side of the low. Consequently, the weather in California will "not be fit for man nor beast," even though the center of the low is off the coast of Oregon. The L1 position shows the low approaching the coast. The L2 position is the low sagging into the Bay area. The L3 position is the low dropping into Los Angeles.

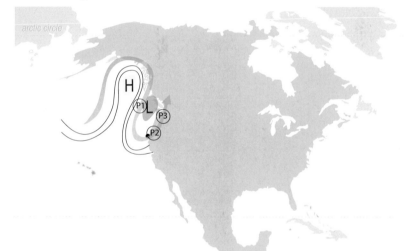

Another old rhyme is useful in these circumstances. It goes, "With your back to the wind, the storm is to your left." This is a very useful rule to apply to any situation in which you would want to track a storm. Let us apply this rule to the various conditions. In P1, if you were on a ship off the coast of Vancouver Island with the wind at your back, the center of the storm would be to your left. The wind here would be telling you that the storm had passed. In P2 if you were in Central California and you put the southwest wind at your back, the center of the low would be to your left. The wind here would be telling you that the storm was about to break on you. In P3, with the wind coming from the south, if you were in Idaho, the low would be to the west. The wind here would be telling you to pay attention for future stormy weather.

This insight about the wind, leads to the old sailor's rhymes that allowed them to avoid trouble. Coupled with a bit of cloud wisdom, this rule is very reliable, whether at sea or on land. It goes, "A backing wind [counterclockwise] brings weather by; a veering wind [clockwise] will clear the sky." *Veering* is a nautical term for clockwise, while *backing* means counterclockwise. These terms refer to the direction of the wind as it shifts around your position as a storm approaches.

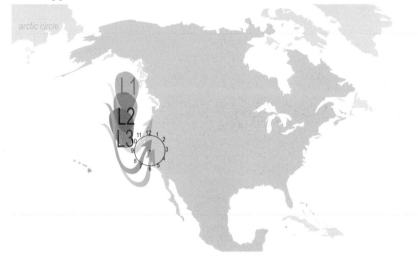

A backing wind would move from 9 to 6 on a clock face as a storm moved towards your position. Suppose you were on the West Coast in the vicinity of Sacramento. In figure 6, we see the storm coming down the coast with the face of a clock on Sacramento. With the storm to the northwest, the winds at the leading edge of the dangerous semicircle would come into Sacramento from the west—that is, 9 o'clock to your position. As the storm came further south, the wind would shift to the southwest—that is, it would come from a 7 o'clock position in Sacramento. As the storm hit, the winds would be from the south, or the 6 o'clock position in Sacramento. In this sequence, the wind would have backed through the clock backward as the storm approached your position on a track from northwest to southeast. This is a backing wind. As the storm passes, the winds would follow the clock in a clockwise fashion as the skies cleared.

Putting these ideas together to form a protocol for predicting a storm, one might see something like the following sequence.

Ring around the Sun, or "sun dogs"

If in the early morning the flies are biting or sitting on your elbows, you can assume that the barometer is about to start falling. You remember the rhyme: "Flies on your elbows, rain on your head." This prompts a person to go out and check the wind. If it is from the west, one would scan the skies to look for clouds that might tell of a change.

If, when you went to look for clouds, you were to see a ring around the sun (above), it would be prudent to start watching the wind for any backing movement (moving counterclockwise in the position from which it is blowing). "A backing wind brings weather by."

If, by mid-morning, the wind was backing from the west toward the southwest, and a mass of "cloud streets" (next page top) have formed in the western sky, one would know that the upper air is the site of turbulence. Cloud streets are ripples in the upper atmosphere that speak of rising warm air currents that lift skyward before a storm. Seeing these, one would want to know the general direction from which the low is coming. One could predict, by the rule of the wind at your back, that a storm is coming from somewhere to the northwest. "With the wind at your back, the storm is to your left."

Cloud streets (top); altocumulus (bottom)

Nimbus

The task would be to determine how far away it is and whether it will come toward your position.

If the wind has shifted to the south by the late afternoon, and the cloud streets have lowered into altocumulus, then you would know that the sky is "lowering" toward your position. This would indicate that the cloud types are getting lower and that the front around the storm is approaching your position. As noted, altocumulus clouds are often called mouton, or "sheep." The old rule was, "When sheep flock together, beware of foul weather."

If these signals are present and the wind backed farther from the south to the southeast, then you can know that you will be in for a blow. If the wind holds steady from the south and does not shift to the southeast, you would know that you are not in the dangerous semicircle. If the wind moves from the west to the southwest and holds steady, you might escape the storm as it passes to the north.

In any case, the sky will probably turn to rain, or nimbus, types of clouds (above). These are the lowest types, and they indicate that the front is passing

Mackerel sky

your position. Rain comes from these clouds. They look dark, torn, and ragged and move rapidly across the sky because they are so close to the ground. They usually remain in the sky for a few days.

As the storm is passing, look for a mackerel sky (above). This high cloud often tells of a change, as the old rhyme says: "Mackerel sky, mackerel sky, dry turns wet, and wet turns dry." If these cloud forms are accompanied by a shift in the wind veering from south through the southwest to the west, the weather will brighten and the wind will be calmer. This wind moving with the clock would be a veering wind, going from 6 o'clock to 9 and to 11. Winds out of the west/northwest would then signal a change to fair weather with a calm and clear night. For it is said, "The west wind, like an honest man, goes to bed at sundown."

North American Climate Patterns

The term *climatology* refers to the default level of weather for a particular geographical area. This means that, if all signals fail to give a clear weather picture of a coming time period, the forecaster will refer to the historical climatic mean for that area on that date in order to make a prediction. In weather bulletins in which the forecast is not really evident from an analysis of the data, the term *climatology* will be printed over the chart as an indication of the default. To make forecasts, climatologists take the most prevalent pattern for that date as the basis for seeing into the future. If it rained for the past six years on the first of March, statistically it will probably rain on the first of March this year. This is the basis for climatology. Many long-range forecasts are made using such climatic techniques to make the predictions. At the foundation of this are several semi-permanent climatic scenarios that can be predicted in advance based on the climatic records.

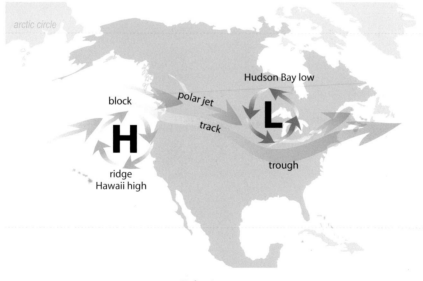

Polar jet stream

The climatology of North America is created by two fundamental atmospheric patterns. In the language of climatology, one pattern is known as a "ridge." A ridge is a mass of high-pressure air that serves to steer the jet stream into paths known as "storm tracks," or simply "tracks." A ridge that persists for more than a few days is known as a "block." High-pressure ridges that form blocks tend to be situated in predictable places in a given season, owing to certain climate patterns. The fundamental task of the forecaster is to be able to determine, in advance, the position of the blocking ridges that will steer storms during a particular time frame.

The opposite pattern is known as a "trough." A trough usually forms in the lee of a ridge. It is a place where storm fronts interact. Blocks generally remain stationary, while troughs migrate west to east, carrying storms with them. Storms steer, or track, around blocks, and the climate patterns on the continent are the result of blocks being placed in particular positions at particular times as the storms track around them.

The image (page 31) shows a stream of rapidly moving air high aloft, known as the polar jet stream, moving across the top of a ridge of high-pressure air in the eastern Pacific. The polar jet stream is the primary weather maker for the Northern Hemisphere. It is a river of high-latitude, fast-moving air at a high altitude. The ridge in the image is a fundamental block for North America known as the "Hawaii high." During summer, this block expands across the whole eastern Pacific and dominates wind patterns on the West Coast by expanding up into the latitude of Oregon. This brings drought conditions to California and Oregon during the entire summer, since storms cannot get past the block and move south into the coast.

The polar jet stream in the image is coming off the top of the Hawaii high and moving from west to east across the continent. As it enters the continent, the polar jet stream is being deflected north by the mountains in the western states. Then, in the lee of the mountains, a trough is formed in which the jet stream dips first south over the Midwest and then back into the northeast as it exits the continental U.S. This is a "lee trough." This overall flow is the most fundamental pattern for North American climatology. It brings cold Canadian air into the U.S. from the north. In summer it can also draw up warm air from the south for the areas east of the Mississippi. In the West, the mountains create a barrier for Pacific air, preventing the easy entry of Pacific moisture into the western U.S. for most of the year.

Troughs, or low-pressure areas, are formed by rising warm air. Winds blow into the center and upward in a trough as they also circulate counterclockwise.

A high-pressure area, by contrast, is formed by falling cool air. Winds blow down and out from the center around a ridge as they circulate clockwise. Lows and highs can sometimes generate semi-permanent troughs and ridges respectively. These features fall out of the general circulation and can persist over weeks in the same place without moving to the east. Semi- permanent features are the source of most climate patterns.

The Hawaii high is only one of the possible blocking patterns in the atmosphere. On the West Coast, there is also the Pacific/North American pattern, the Mississippi ridge, the Bermuda high, the Greenland block (or North Atlantic Oscillation), and the Azores high. These blocks arise in different places and for different reasons, but they all share in common the fact that they are the single greatest influence on climate patterns in North America.

Each block is set opposite a trough, or low-pressure center, that is linked to the expansion and contraction of the associated block. The Hawaii high has the Aleutian low as a polarity. The Mississippi ridge has the Hudson Bay low as a counterpoint. The Bermuda high has the Greenland block as its opposite. In the North Atlantic Oscillation, the Iceland low is counter to the Azores high. As the polar jet weaves between the different permutations of blocks, certain favored tracks for storms arise. The relationships between blocks and tracks and the climate patterns that result from them will be detailed in the following chapter.

When a continental high over the western states is very strong, the lee trough on the eastern side of the Rockies is sometimes strong enough to form a stationary low-pressure area that is placed far to the east of the Rockies, just south of Hudson Bay (next page top). This placement and the usual counterclockwise circulation around the Hudson Bay low would bring stormy weather and cold temperatures into the eastern U.S. In this pattern, it would be dry in the High Plains, and during winter cold could penetrate down into Florida, threatening citrus crops.

When there is a strong high-pressure ridge over the southeastern U.S., it tends to push the jet stream to the north, causing the cool polar jet to move horizontally across Canada. When that happens, warm air flows into the continent from the south in the form of a monsoon jet (next page bottom). In this situation, moisture from the Gulf of Mexico would most likely rise

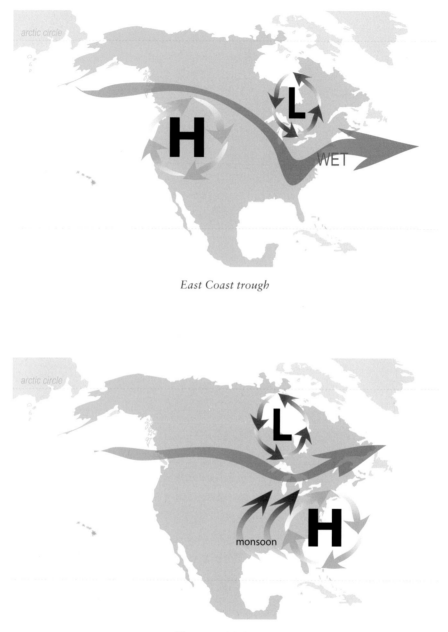

East Coast trough

Monsoon jet stream

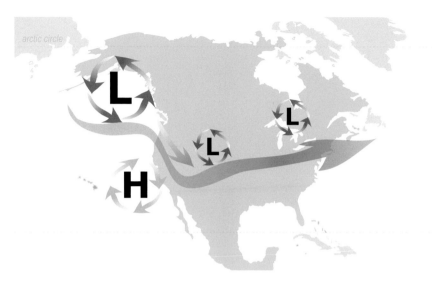

Gulf of Alaska pattern

into the Midwest. As the cold air from the north meets the warm air over the Great Lakes area, it could be expected that strong summer thunderstorms would be the result for the Great Lakes and the northeast. From this we can see that the geography of the U.S., especially the placement of the mountains on a north–south axis in the west, creates a particular physically driven dynamic for weather on the continental U.S.

Weather patterns move from west to east in the flow of the jet stream. Because of this, the weather in the United States has its origin in the Gulf of Alaska (above). The climatology of that area is such that there is a consistent low-pressure area known as the "Aleutian low," which forms in the western Gulf of Alaska. The Aleutian low is linked harmonically to another low over Hudson Bay area. When the Aleutian low is strong, the Hudson Bay low is weak, and vice versa. With a strong Aleutian low in place, the Hawaii high is compressed and the cold polar jet enters the continent over the Desert Southwest in a pattern known as an "inside slider." Fronts slide down the continental side of the Sierras, putting storm energies into the intermountain west, where a low typically forms in the vicinity of Denver, with storms and cold moving into the central High Plains. A pattern such as the inside slider can often bring cold and snow into the mountains of Utah and even into California, threatening frost-sensitive agriculture there. In general,

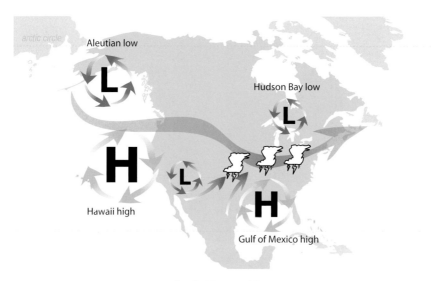

April, May, and June

this pattern includes a strong low over the Aleutian islands or the Gulf of Alaska and a reciprocal weak high over Hawaii, as well as a weak low over Hudson Bay.

Taken together, these relationships can give rise to several different patterns of climate in any given year. The following synopses are seasonal patterns that form the foundation for climate scenarios in North America. In this chart (above), we see the annual spring climatology for April, May, and June. In the eastern Pacific, there are two general features in the spring and early summer. One is strong high-pressure air between Hawaii and the coast. The high pressure forms a ridge-like air mass of clockwise moving air that effectively blocks the polar jet from dropping south into the mid-latitudes. For this reason, every year there is a drought condition on the West Coast from May or June until the Hawaii high diminishes in September. This rotating air mass forms a semi-permanent high-pressure ridge that lodges against the Sierras and keeps skies clear and a predominating northwest wind blowing into the coast during most of the summer. This in turn sends warm surface water to the south to be replaced by cold deep ocean upwelling currents in a narrow strip against the coast of California. It is this cold upwelling that is being driven by the relentless clockwise winds and create conditions for the famous San Francisco fogs during the summer months. The Hawaii high builds up to summer strength during the months of April, May, and June. When this happens, the center of storm activity in the Aleutian Islands is reduced in prominence. Polar breakouts of very cold air are much less frequent or severe during the summer months, since northern latitudes are being warmed by the midnight sun every day of summer.

As a result, another annual climatic east Pacific pattern is the reciprocal, or response, to the growth of the Hawaii high, this being the shrinkage of the Aleutian low in March and April. However, as the Hawaii high builds in strength against the Sierras, a low in the Desert Southwest builds in response and begins to create storm centers as the desert heats up. Convection currents set the stage for lines of thunderstorms streaming to the east along a path into the mid-Atlantic seaboard. Moisture from the Sea of Cortez and the Gulf of Mexico feeds this storm genesis. At the same time, the polar jet coming from the Gulf of Alaska often dips down into the Great Lakes area and meets the warm monsoons from the southwest, forming vital early spring rains for the High Plains and the Plains states. Typically, a dry high-pressure area forms over the coastal states of the Gulf of Mexico during these months. This high forces the storm jet up into the Midwest at this time to meet the polar jet coming down from Alaska. Between these areas, spring storms arise from the Ozarks to the Great Smoky Mountains, bringing early rains to support the corn and soybean agricultural base of the Midwest. Failed April rains in this sector send the soybean market into a frenzy.

In the following chart (next page), we see that the Hawaii high is firmly established by July and August, that the Aleutian low is usually a non-player for the West Coast, but that the low in the Southwest has grown considerably. This is often the wettest time in the Southwest. This occurs because of the strong heating taking place here during the summer months. The polar jet moves north, away from the Dakotas, bringing a dry period to the northern tier states in June and July. At the same time, a ridge of high-pressure air usually forms over the Mississippi Valley, blocking any fronts coming from the low in the Southwest.

This usually results in a storm line that runs from the Ozarks eastward into the Carolinas and to points north during June and July. Along the northwestern side of the blocking ridge, strong thunderstorms develop that sometimes give birth to tornados. During some years, the ridge is placed

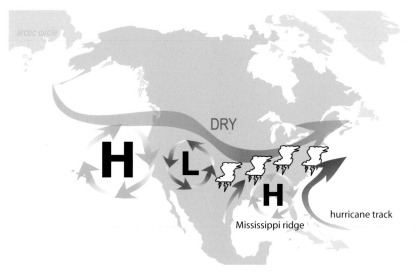

July and August

more to the west, in which case the High Plains gets a monsoon summer of storms and lightning. If the ridge is placed more to the east, then there can be flooding in the Mississippi Valley as lines of thunderstorms dump rain along the sloping line of the blocking high or ridge. As summer progresses, the blocking ridge moves into a north/south placement, paralleling the drainage of the Mississippi. The tornados and thunderstorms also move north into the northern Plains and the Corn Belt. To the east of the block, the very early onset of hurricane season is usually the most active part of the July/August climatology for the Southeast, producing local thunderstorms in summer when the hurricanes curve out to sea, as well as heavy rains when they travel up the coast. Early hurricanes can occur during late summer, although climatically September and October are the true hurricane months.

The early September patterns produce a burst of thunderstorm activity for the upper Midwest as the Mississippi block grows to a maximum and then quickly fades. By the middle of September, the Sun is making rapid progress to the south, the Hawaii high shrinks, and the Aleutian low strengthens. The Sun, rapidly shifting in declination, is causing the heat balance for the northern hemisphere to be transformed. The summer monsoon

September and October

from the Gulf of Mexico collapses. Any residual warm air on the continent flows northward, initiating Indian summer in Wisconsin and Minnesota. By October, the transformation is usually complete, with no thunderstorms in the Midwest. At that time, the strengthening Aleutian low sends the first fall storms into the Pacific Northwest. The monsoon in the Gulf of Mexico breaks down and the Mississippi block fades. This allows the jet stream from the Pacific to begin to be a stronger element in the weather for the continent. On the whole, the Midwest finds this time the most tranquil of the year. On the East Coast, hurricane warnings continue but the tracking usually moves from an early season classic east-to-west track from the Cape Verde islands off Africa and into a shorter track from the Caribbean through the Gulf of Mexico.

The next chart shows (following page) the mean winter climatology. It should be noted that this is extremely simplified, but it is useful for the purpose of illustrating some principles. The Aleutian low is strong, while the Hawaii high is much diminished. The Aleutian Low is the source of most of the storm energies that come into the West Coast during the winter months. Cold storms flow directly south out of the Gulf of Alaska, bringing snow to the Sierras. Wet storms usually drop from the Aleutians

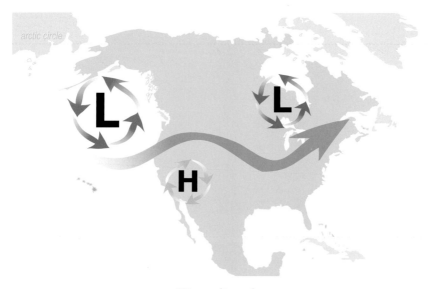

Winter climatology

into the Pacific in the latitude of Hawaii before turning east and on into the coast. There is a delicate balance for these storms in the vicinity of Mt. Shasta at the top of the central valley in California. The winter polar jet often steers to the north of Mt. Shasta when the track straight out of the Gulf of Alaska is active. When the polar jet drops into the vicinity of Hawaii, the storms will track to the south of Mt. Shasta, bringing rain into California. It is often the case that the strength of a high over the Southwest determines which way this will go.

During early winter, there is a high in the Southwest that builds up when the first few cold storms come south out of the Gulf of Alaska, leaving cold air in the Basin and Range area between Nevada and Denver. The Hawaii high and the Southwest high sometimes connect to each other in winter, causing blocks to the south of Mt. Shasta on the West Coast. That scenario results in clear and cool weather for the West Coast.

The Hudson Bay low is the main low-pressure feature of the atmosphere at this time on the continent. This forms when the land surrounding the bay cools much more rapidly than the water does. The water then serves as a source of rising warm air and is the site of a semi-permanent low-pressure area for most of the early winter. The Bermuda high, which is active during hurricane season,

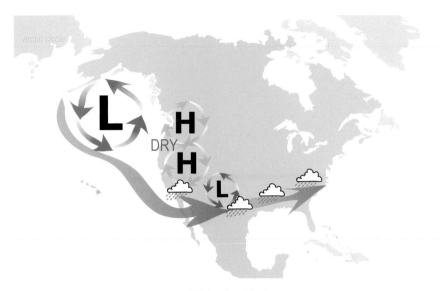

PNA+ (positive)

shifts to the east to settle over the Azores off the west coast of Africa, where it dominates the eastern Atlantic during the winter months.

In addition to these patterns, there are many variations that can greatly influence the weather over periods of weeks to months. Using these fundamentals, we can now look at a few recognized climatic patterns that accent one feature above another feature in a given season, giving rise to climatic variations. Climatologists consider that each scenario tends to fluctuate between a positive phase and a negative phase, giving rise to different climate patterns in different years.

One fundamental climate scenario is the Pacific/North America (PNA) pattern (above). It occurs in both positive and negative modalities. The positive PNA pattern is a variation of the interaction between the Aleutian low and the Hawaii high. It has a strong influence on the weather on the West Coast and in the Southwest. The positive PNA often appears in El Niño years. In this pattern, a high develops along a north–south axis on the West Coast over the mountains from Canada south to Northern California. The center of the high is usually in Alberta, British Columbia, or Vancouver Island. Most often it flows to the north and then wedges into the eastern shore of the Gulf of Alaska. From there it acts to drive the polar jet to the south by blocking

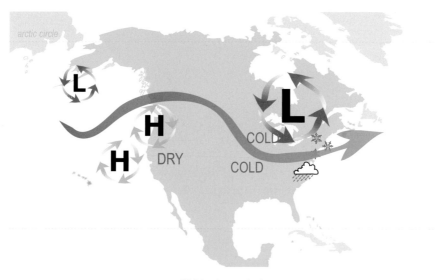

PNA– (negative)

passage into the continent through western Canada. If this block is long-lived, flood conditions are often created to the south as the jet steers in a long curve, or trough, over the waters of the eastern Pacific before making a landfall. Usually in this pattern a cold northern storm is pushed south in the vicinity of Hawaii, and it picks up moisture as it crosses the Pacific. This pattern yields wet conditions in California, Utah, the Desert Southwest, Denver, the Gulf States, and Florida.

In the PNA– (negative) pattern a high from Hawaii forms over the eastern Pacific and connects to a high in Oregon to form a strong block to the polar jet (figure above). As a result the jet is pushed to the north on the West Coast and then descends into the continent at a strong angle. This placement of the highs is often, but not always, the case during La Niña cycles. The PNA– pattern creates severe drought below Mt. Shasta on the West Coast. For the continental U.S. in this scenario, the Hudson Bay low is strong, while the Aleutian low is weak and located to the west. The counterclockwise circulation around the strong semi-permanent low near Hudson Bay dominates the polar jet there and pushes cold, stormy weather down into the continental U.S., usually as a strong trough on the East Coast. Fronts from the Hudson Bay low drop out of Canada on a ten-day interval, bringing snow and ice as far south as Florida.

These patterns are climatic variations on the theme of high and low pressures alternating in various ways in different seasons. It is tempting to think that the polar jet somehow works like a clock or pendulum. If this were the case, forecasting would be simply a case of plotting the statistical mean for a given year and then checking it against the climate record and making a prediction. However, the placement of the low in the Aleutians is tied intimately to the placement and strength of the low over Hudson Bay in a given year. The placement of the Hudson Bay low in a given year is dependent on the placement of the Iceland low. This in turn is dependent on the strength and placement of the high-pressure area in the Azores. The links between these remote air masses are harmonically sensitive and known to climatologists as "tele-connections." These stretch over thousands of miles and are connected by the fluctuations of the polar jet in response to seasonal changes and other more subtle influences. Tele-connections are driven by changes in high-pressure blocking ridges, which are semi-permanent features of the atmosphere of the northern hemisphere.

There are several major low-latitude semi-permanent high-pressure areas in the northern hemisphere. These air masses steer the polar jet into its sinuous path through the temperate regions. The first is in Hawaii; the second is over the Azores; and the third is over Siberia during winter. However, during winter a secondary high-pressure area also forms over the Desert Southwest or over the Southeast, as we have seen. It is the position of this ridge that provides a very good distant warning area for upcoming weather changes on the European west coast. The reason for this is a pattern known as the North Atlantic Oscillation (NAO), which is presented in the next image. A fourth position, which is not as strong or persistent as these three, is the Bermuda high that sometimes forms off the southeast coast of the United States.

In the NAO, a reciprocating relationship similar to the Hawaii high and the Aleutian low exists in the eastern Atlantic. The low is centered in the latitude of Iceland. The reciprocating high is centered in the latitude of the Azores. The NAO+ (positive) is depicted in this chart (next page), in which we can see that, on the American continent, there is a situation in which the Hudson Bay low is strong and is tele-connected across Greenland to a weak Iceland low. The Southeast high is tele-connected to a strong Azores

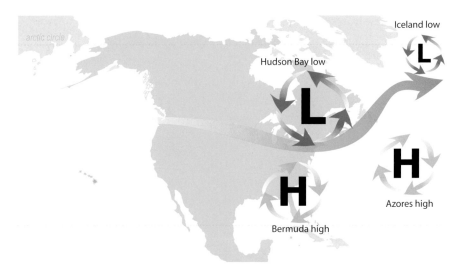

North Atlantic Oscillation NAO+ (positive)

high that has moved to the central Atlantic. The Hudson Bay low is situated between the two highs like a roller gear. This happens when the Hudson Bay low is placed to the east, near the Maritime Provinces. This pattern creates a situation in which Pacific maritime air comes in through Washington State into Alberta, and then continues horizontally across the northern tier and exiting the continent without drawing strong cold down from Canada. The jet moves out to sea and crosses the Atlantic by being shunted along by the circulations around the Azores high and the Iceland low. This makes for strong westerlies across the Atlantic. As the name implies, there is often an oscillation to this pattern. The oscillations can occur over years, weeks, or even days. This is the positive NAO in the U.S.

In the opposite scenario, if the Azores high is strong and more to the west, the center of the high can sometimes shift farther to the west (facing page). If it is very strong, it surges to a high latitude and moves into the area near Greenland. In response, the Hudson Bay low will drift to the west, allowing high pressure into eastern Canada and the Maritime Provinces. This puts a blocking high over the Maritimes. This pattern is known as the "Greenland block," or the negative NAO. There is a flipping of the relationships found in the positive NAO. Such a pattern would bring strong cold and snow into the East Coast of the U.S. It would also warp the polar jet over the eastern

NAO– (negative)

Atlantic, giving rise to storm cycles in the U.K., since whenever the high pressure is strong over Greenland in the west, the area near Iceland tends to deepen in low pressure. In the negative NAO pattern, the low pressure in Hudson Bay and Maritime provinces is replaced by a persistent high-pressure ridge. This ridge blocks the normal west-to-east passage of cold air across Canada and drives it to the south and into the Great Plains and to the eastern seaboard. This results in record cold winters in these areas.

The period of oscillation between these two phases is very erratic and provocative in its implications. During the 1920s, the positive phase dominated. During the early 1930s, the negative phase dominated. In the early 1960s, the negative phase dominated. During the 1970s and 1990s, the positive phase dominated. Correlations between eclipse positions, planetary positions and these dates have been established through the techniques of the eclipse grid model, which will be presented later in this book.

Another strong climate pattern for the continental U.S. is of interest to those who follow corn and soybean crops. This pattern is the low-level jet (LLJ). During the summer, the grain crops in the Corn Belt require regular thunderstorms to keep growing. These thunderstorms arise from a combination of a monsoon-like moist flow of warm air from the Gulf of Mexico interacting with a cold front from the north. When these patterns occur on average

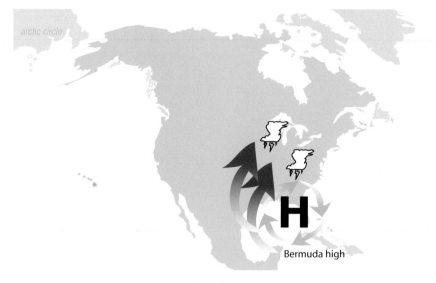

Low-level jet

of a week or so apart, then the crops do well. If the monsoon fails, then the crops fail along with it. The failure of the monsoon, leading to drought and crop failure or the onset of flood conditions leading to crop failure, are both controlled by the LLJ pattern (above).

This feature of the atmosphere is controlled by the placement of the Bermuda high, which is a semi-permanent high-pressure area that oscillates from a position over the central Atlantic in some years to a position against the coast of North America in others. When the high is against the coast, the clockwise circulation of winds around it moves into the Gulf of Mexico and drives the moist air from the Gulf into the Midwest. When this happens, a low-level (around 5,000 ft. to 10,000 ft.) tropical jet stream (red arrows) establishes itself in June or July and brings moisture northward. The moist flow from this low-level tropical jet supports abundant rains. In 1993, this flow created the record floods on the Mississippi and its tributaries. The position of the Bermuda high up against the Gulf coast is the trigger for the low-level jet and is a flood signal for the Midwest.

The pattern opposite to the low-level jet leaves the Midwest in drought conditions. The chart shows the typical placements for the major drought years in the Midwest for 1953 and 1988. The pattern here is one in which a

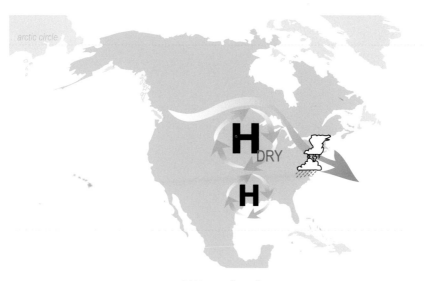

Midwest drought

strong western Atlantic high is replaced by a consistent East Coast trough, whereas the strong high-pressure area is found over the Gulf of Mexico and Western North America. We can see that the low-level jet pattern for the Gulf of Mexico is effectively neutralized and that the dry high-pressure area over Denver steers the Pacific jet stream up into western Canada and then down into the East Coast to form the rain trough there. Most Midwest droughts feature a ridge like this one. This is the type of pattern that set the stage for the Dust Bowl of the 1930s. The Dust Bowl will be covered in detail in a later chapter of this book.

Winter Storm Tracks

The placement and intensity of blocks create many variations in the formation of the jet stream loops that create the storm tracks across the North American Continent. This is especially true in winter, when the Hawaii high can maintain enough strength to form a cushion across the eastern Pacific but cannot block storms from the coast. This often occurs when the center of the high is well off the coast. At such times, the storm jet can ride along the northern edge of the cushion and steer cold Aleutian storms from the north, southward along the coast.

This flow brings intense cold into the central valley of California on the eastern side of the block. This type of pattern happens at approximately ten-year intervals, destroying orange and lemon trees in California's Central Valley. The presence of strong cold into the West Coast is the Alaska track.

When a block is situated right on the coast itself or over the inter-mountain areas, the jet stream rises into Canada and then drops into the High Plains or the Midwest. This is the central track. In some winters, the central track is the prevailing storm track. If the block is situated farther east, over the Great Lakes, the Canadian cold drops into the area east of the Great Lakes. This is the eastern track. This is often the source of unusual cold snaps in the Northeast. If the cold outbreaks are strong enough, they can penetrate into the south and the Gulf Coast, threatening the Florida orange crops.

Looking at these blocks in the long term, sometimes it can be seen that a dominant or prevailing block during one season moves from an eastern position over the Midwest one year, into a more westerly position over the Rockies and the next year, and shifts over the West Coast the year after. This phenomenon points to larger rhythmical patterns that are of interest to the long-range forecaster.

A WORD ON ZONES:
CLIMATE ZONES IN NORTH AMERICA

The climatic zones in this work are arranged so that they describe the most general patterns of storms in the North American continent. These continental patterns are related to the two jet stream gates on either end of the continent. The four zones that involve the gates are the West Coast, Mountain/High Plains, Mid-Continent, and East Coast. This division is seen in the next image. The Southwest and the Gulf Coast are not influenced much by the two gates. They are influenced much more by the patterns of easterly winds around the Gulf of Mexico. Since the weather in those zones do not rely to a great extent on the polar jet stream incursions into the continent, they are not the focus of this chapter.

The West Coast zone includes Washington State, Oregon, and California. The area between the Columbia River gorge near Portland in the south and, in the north, the Straits of Juan De Fuca on the southern side of Vancouver Island, British Columbia, is the focus area for storms moving across the Pacific. We

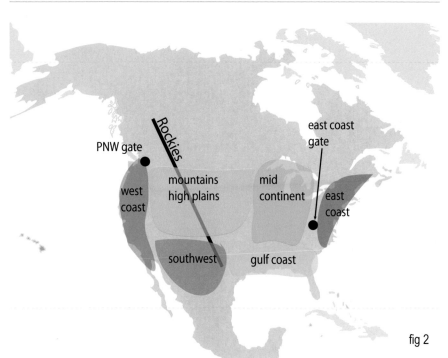

fig 2

could call this geographical area the Pacific Northwest (PNW) gate. On the West Coast, most storms that appear in the eastern Pacific flow through the PNW gate following the jet stream. The jet stream is a river of high-altitude air currents that steers storms across the Pacific Ocean. The three states on the West Coast are said to have a maritime climate because their weather patterns are so strongly influenced by the Pacific Ocean.

The next zone to the east of the West Coast zone is a bit more difficult to characterize. It is possible to link what is often called the Basin and Range area, with the area known as the High Plains. The Basin and Range states from west to east are, Nevada, Idaho, Utah, Arizona, Montana, Wyoming, Colorado, and New Mexico. The climate in these states is generally arid with little rainfall and a typically very cold winter. The dryness is influenced by the fact that many different ranges of mountains cover the area between the Sierras and Cascades on the West Coast (approximately 128° w longitude), and the Rockies at Denver (105° w longitude) in the east. The width of the great geological basin between these two ranges is about 800 to 900 miles. This whole area is filled, north to south, with mountain ranges running north and south, which prevent the easy eastward passage of Pacific storms. Storms from the Pacific tend to avoid going over these ranges and instead seek mountain passes running east and west through the ranges. As we saw earlier, most Pacific storms, and hence most rain, track toward the wide pass at the PNW gate. This is why the western states are so dry.

From their farthest eastward position at Denver, the Rocky Mountains swing to the northwest into western Montana and continue northwest into the west coast of British Columbia. In western Canada, the Canadian Rockies are found at 120° w longitude. Thus, the mountains swing from a maximum eastward position in Denver on a line to the northwest (black line, previous image). This tracking line shifts their placement to the west approximately 800 miles of longitude as they move toward the north. As they swing to the northwest, the Rockies come very close to meeting the PNW gate mentioned earlier.

Just to the east of the gate, there is a high valley in eastern Washington State, formed by the drainage of many great rivers, such as the Snake, Yakima, Spokane, Columbia (image right). The city of Spokane sits at the top, or northeastern end, of this large valley. The mountains to the north and south of Spokane are higher than the mountains to the east of the city. With a broad valley to the west, and peaks to the north and south, these geographical features create a west-to-east pass through the Rockies. This pass acts as a compression release valve, allowing storms from the whole Pacific to pass from the PNW gate and funnel east through Spokane, and then through the Rockies east of Spokane with the minimum of resistance. This pass steers Pacific storms into northern Montana and southern Alberta as they move east on the jet stream.

Once the storms are through the PNW gate and then past Spokane, they re-form, usually in the Alberta area, and drop into the High Plains of eastern Montana and the Dakotas. However, they rarely drop due south into the High Plains, and they rarely have the moisture in Montana that they develop farther south and east in the Great Plains, when the southbound cold storm fronts encounter moist, warm air from the Gulf of Mexico over the western states of the Midwest. This storm track often bypasses the High Plains in Montana and Colorado, forming an arid rain shadow in the lee of the Rockies over the High Plains. Consequently, in this work the High Plains are included in the same arid climate zone as the high deserts of the western states. The High Plains area includes eastern Montana, eastern Wyoming, Eastern Colorado, the western Dakotas, western Nebraska, western Kansas, western Oklahoma, and northern Texas. The High Plains zone shares soil types and the general patterns of rainfall and jet stream motion with the majority of the central and northern mountain states. Since it is so dry, the High Plains can be farmed only by pumping fossil water for irrigation.

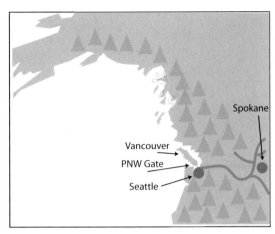

PNW gate

This area is sometimes called the Wheat Belt. In this work, the combined Basin and range area and High Plains zones are called the Mountain/High Plains zone.

The Midwest or Mid-continental zone is much more moist than the Mountain/High Plains. The true Midwest or Mid-continental weather zone does not begin to take effect until the drainage of the Missouri River is encountered by storms migrating from Alberta into the Dakotas ("Midwest" image, next page). In this work, the dividing line is placed vertically at approximately 100° w longitude, along a line drawn between Fort Worth, Texas; Wichita, Kansas; and drifting to Rapid City, South Dakota, and the Black Hills. This line is the beginning, or western edge, of the Midwest and the termination, or eastern edge, of the Wheat Belt.

The Mid-continent zone includes the central states, the eastern Plains, many of the Gulf States, and the area between Minneapolis and the eastern mountains in the North. This area includes southern Minnesota, Iowa, Indiana, Illinois, Wisconsin, and Ohio and is commonly called the "Corn Belt," because there is enough rain from thunderstorms each summer to support the growth of corn and soybeans. The summer thunderstorms are caused by the meeting of moist air from the Gulf of Mexico and cold air either coming through the gate in the Pacific Northwest or dropping down from the Canadian Plains. Typically, a cold Canadian front flows in a

Midwest

southeasterly direction toward the Missouri River in the Dakotas (left fork on the chart). Most typically, storms track the path of the Missouri River into St. Louis, where it joins the Mississippi near the junction of eastern Missouri and southern Illinois. There, the Canadian cold air encounters moist, warm air from Texas, Arkansas, and Louisiana. Thunderstorms erupt and their abundant rains encourage the lush growth of the moisture-loving corn and soybean crops.

To the east of the Mid-continental zone is the East Coast zone (next page). The mountains on the East Coast tend to steer most Mid-continent storms into the northeast. There is a gap in the eastern mountains along the southern border of Pennsylvania. The mountains to the north in Pennsylvania and to the south in West Virginia are higher than the mountains in this border area. This gap creates a storm gate. As a result, many storm centers traveling across the continent surge toward this area and spread out again into the Boston–Washington corridor. This East Coast gate in the north has a counterpart in the south, where the mountains in eastern Tennessee reach a maximum height and then suddenly fall to the Gulf Coast Plains near Chattanooga. Storms coming from the Gulf Coast often use the area between Chattanooga and Greenville, South Carolina, as a passage to the East Coast.

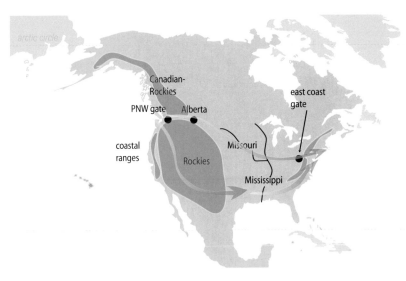

East Coast gates

The abundant rain from these storms supports the lush rain forests in the upland areas of the Carolinas. If a winter Gulf Coast storm is strong, it can sometimes veer to the north as it flows around the eastern slopes of the southern mountains and eventually reach the northeast by coming up the coast as a nor'easter. The other great influence on this area comes in late summer and early fall in the form of landfall hurricanes. Cape Hatteras, on the easternmost tip of North Carolina, is the most preferred landing spot for hurricanes coming from the African west coast, although hurricanes can make landfall anywhere along the entire East Coast and Gulf Coast, depending on the currents in the atmosphere

READING THE WINDS ON THE CONTINENT

A fundamental skill for forecasters is reading the winds. Shifts in the direction of winds surrounding an approaching storm reveal the movements at the centers of the large air masses driving the storm. The positions of the prevailing winds in a given location during longer time frames reveal longer-lived storm patterns that form the basis for climate study. The longer seasonal patterns of winds on the continent are best studied by using a wind rose. The wind rose indicates the prevailing positions from which the wind will blow

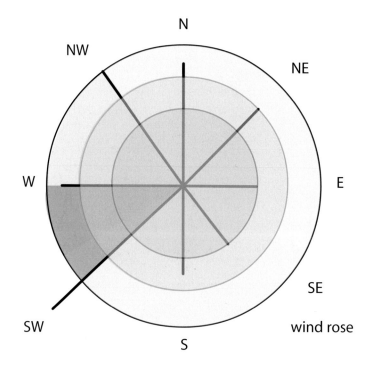

throughout the year in a given location. The chart above shows the wind rose for Philadelphia, where the most common prevailing wind at all times of the year is from the southwest. The wind rose shows this in the longer line that extends to the southwest beyond the outer circle. It is longer than any other line in the diagram. The next prevailing direction is from the northwest. This line just reaches the outer circle. The next prevailing directions are due west and due north. These lines reach just beyond the second circle. This is followed by the northeast line, which reaches the second circle. The rest of the directions for this location are not prevailing during the year.

To understand this wind rose it is useful to look at the mean tracks of storms across the Midwest into the Philadelphia area during the storm season. In the next image (opposite), we see that, no matter where one of the continental storm tracks has its origin, it tracks through Philadelphia (black dot) from approximately the same direction. The Great Lakes track across the northern tier of states, brings warm maritime air from the Pacific zonally or horizontally across the top of the U.S. during the winter. Storms ("L" in the chart) in the northern hemisphere have a counterclockwise circulation. A

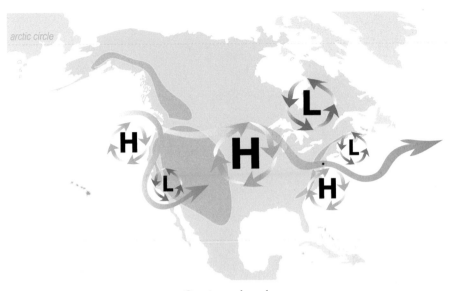

Continental tracks

storm tracking into Philadelphia would have a circulation that would place the prevailing wind to the southwest as the storm approached Philadelphia from the northwest. This is depicted in the upper track.

A storm approaching Philadelphia from the lower Midwest would also produce a southwest wind as it came into the area. This is depicted in the lower track. The circulation around a transiting low-pressure area will predominantly bring southwest winds into Philadelphia no matter what direction it comes from across the continent. All storm tracks lead to the northeast corridor. To see why this is so it is useful to look at the dominant high and low pressure areas and their circulations during the winter months.

In this next image (following page), the large low-pressure area over Hudson Bay is the dominant feature of the continental circulation during the winter. This is because the ground around this enormous body of water is much colder than the water. Water retains its heat much longer than land. As a result, during the winter the Hudson Bay area is a consistent source of rising air. Rising air produces low-pressure. The low-pressure circulation is counterclockwise so the prevailing circulation around this semi permanent low is from the northwest in the Philadelphia area (short red arrow). This gives us insight into the northwest element of the Philadelphia wind rose.

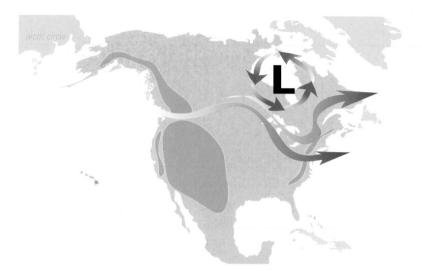

Hudson Bay low

The circulation is so immense around this low in the winter that it even pulls air from the Pacific Ocean into the continent when it is very active (long red arrow). This maritime air, as we have seen, will steer lows across the northern states and then funnel them into the northeast giving rise to the prevailing southwest winds as the fronts pass the east coast to the north of Philadelphia.

Occasionally, the Hudson Bay low is influenced from the east by the presence of a strong high-pressure area over the Maritime Provinces or Greenland. This high-pressure area, or block, is known as the "Greenland high"(opposite page). A block here prevents the passage of the warm maritime air across the northern states. The Hudson Bay low starts to pull cold air down from the western and central Canadian plains. The cold air pushes the storm jet down into the High Plains states. When this happens, storms track south, farther to the west, and penetrate deeper into the south on the continent. Usually, when the Greenland high is active, a trough forms over the East Coast. Storms track across the Central States and then, as they approach the East Coast, they run into the trough.

We can see in this next figure (opposite) that the eastern leg of the trough roughly parallels the East Coast. This is due to the fact that the high-pressure from Greenland often blankets the whole western Atlantic during these events,

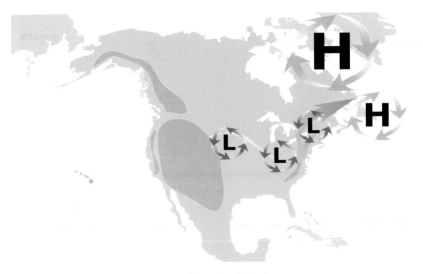

Greenland block

making the northward-moving jet run parallel to the East Coast. The trough usually settles against the offshore high and has little incentive to move east across the high-pressure ridge. In the chart above, we see a small low traveling into the East Coast around the lower leg of the trough. The circulation of the small low brings the prevailing wind to the southeast as the low approaches Philadelphia and then from the northeast as the storm passes by.

In the next image, however (following page top), we see the situation when the trough is placed so that the eastern leg is out over the ocean. This happens when the block is strong to the north but weaker to the southeast over the ocean. The trough drifts to the east. The chart shows a detail of a low moving east along the lower leg of the trough, then turning north over the ocean. The circulation around the low puts the wind out over the ocean on the leading edge of the transiting low. The wind then shifts to the northeast as the transiting low moves up the coast approaching Philadelphia from the south. This is the famed nor'easter that brings wet snows and blizzard conditions to the New England and northeastern sections of the eastern seaboard. It also gives an insight into the strong elements in the wind rose from the northeast.

In the next image (following page bottom), the Greenland block is placed into a context of the whole Atlantic Ocean. The placement of the

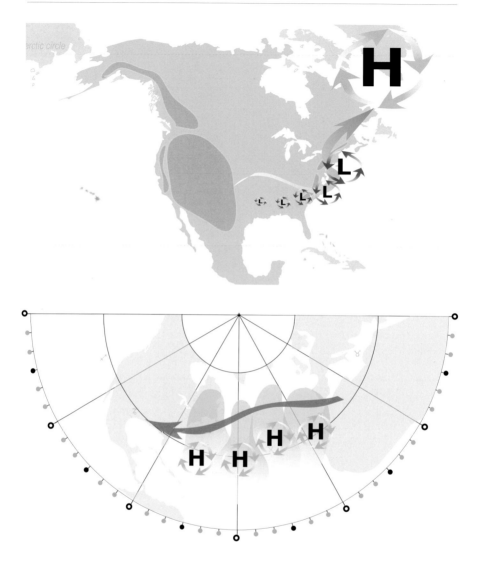

Nor'easter (above); Atlantic oscillation (below)

Greenland block far to the west over the Maritime Provinces is, in certain years, a regular event during the winter. The usual pattern is that the high that is normally over the Azores off the west coast of Africa, surges to the north into the north Atlantic. The high latitude high-pressure ridge then moves westward over the next ten days. In certain seasons this pattern can oscillate a few times a month. For a few days the high is over

Iceland then it starts to drift to the west at a high latitude ending up over Greenland at the end of the week. The ridge can oscillate between Iceland and Greenland for several weeks. When that happens, at about ten-day intervals, the continental jet stream on the U.S. is pushed back toward the west as the Hudson Bay low elongates and streams to the south because of the Greenland block. The Greenland block may stay in position for a week or so and then the whole cycle starts again near Iceland. When the block is in place over Greenland, the polar jet brings cold into the eastern seaboard. When the block is not in place the polar jet pulls air off the continental U.S. and shunts it out to sea. These patterns are the major weather makers for the eastern third of the country during the winter.

The upper layers of the Earth's atmosphere leading to space
(NASA Earth Observatory)

CHAPTER 2

THE TWO-DAY SURGE

Most scientific research starts with the formation of an idea or hypothesis about how a proposed experiment will turn out. Then the experiment is set up either to prove or to disprove the idea or hypothesis. As an approach to chemistry and physics, the hypothetical method has no equal. In the realm of life sciences, however, the type of experiments required by the hypothetical method sometimes have difficulty revealing the more subtle interrelationships among phenomena. In meteorology, this is evident in the vacillating longer-range forecasts generated by the various numerical models. The vacillation in the forecast occurs as each computer run iterates the results of the previous run into a new run. The hypothesis in this case is the programming used to model the process. In the short run, the programmer has to work from a program modeled on the consideration that the forces supporting today's weather will be the same forces supporting tomorrow's weather. In fact, there is a robust statistical basis for such a forecast. The problem arises, however, when such short-range methods are applied to long-range forecasts. In the long term, the weather two weeks from now will probably not resemble today's forecast. This is because the models struggle to describe phenomena extended over thousands of miles rather than describing phenomena that are a few days upstream. As a result, computer modeling for accurate long-range forecasting is still a thing of the future.

In contrast to the hypothetical method and numerical modeling approach, there is another approach to science known as "phenomenology," the systematic exploration of how phenomena unfold in time. When a phenomenon emerges from the general background of forces, it follows patterns that are understood only when the overall movements of the field are recognized. Once the field is recognized, the future emergence and subsequent evolution of the phenomenon can be predicted. The trick is to refrain from forming a hypothesis until a perception of the whole field arises. In this way, the phenomenological approach to science is the opposite of cause-and-effect relationships, which are the foundation of the conventional physical sciences.

To form a phenomenological experiment, an observation is made and recorded. Then another observation and still another observation is made until the field of observations attains a certain critical ripeness. At that stage, the data are reviewed as one looks for what mathematicians call "strange attraction." This means that parts of the phenomenon moved in similar ways in similar time frames. It is called "strange," because we do not know why synchronous attraction occurred. Nonetheless, it is enough for a phenomenologist to recognize that attraction exists, because the goal is not to prove a cause-and-effect relationship but to glimpse the largest movements of the data field. Eventually, all roads lead to Rome, and practical insights into the details of the relationships can be observed and understood once the whole is perceived. This can happen even when there is no understanding of the cause-and-effect relationships in the details that compose the field. However, these details often emerge naturally at the end of the process rather than determining how the experiment is set up in order to prove the hypothesis before the experiment is undertaken.

In a broader sense, phenomenology is the study of patterns that can help the observer to grasp the absolute, logical, ontological, or metaphysical Spirit that is behind the appearance of the phenomenon. This type of study was pioneered by the great German poet and scientific researcher Johann Wolfgang von Goethe, who believed that any phenomenon reveals its pattern of "becoming" to the observer when the observer pays attention to the rhythmic cadences perceptible in the sequences leading up to the actual appearance. It is through careful and systematic observation of the rhythmic elements of a phenomenon as it unfolds in time that the more subtle aspects of the relationships within the whole context of the phenomenon can be understood. While in the standard experimental method the essential is to home in on differences, the essence of the phenomenological method is to look for attraction, synchronicity, or similarity. That means that when two events show synchronicity in time they may be considered a symptom in that they are part of a larger whole that is not yet perceived. Synchronicity of timing in rhythmic events eventually points to the larger motions of the whole data field. The experimental protocol is to keep looking for instances of attraction, synchronicity, or similarity until a larger perspective is revealed.

The following studies were arranged as a phenomenology of how patterns in the motions of the Moon and other planets are synchronous with perceptible movements of large air masses in the atmosphere. To do this, it was necessary to find a technique of projection that allowed for tracking the lunar motion over specific locations on Earth. This technique was realized after fifteen years of systematic phenomenal observations of the synchronicity between the movements of the Moon and the accompanying movements of blocking ridges in the Northern Hemisphere at the 500mb level. The 500mb level is approximately half way up in the troposphere, the place where weather systems develop. The 500mb level is a very sensitive layer that is quick to respond to lunar movements, as the following studies will show. It should be emphasized that the attraction between the planetary movements and the atmospheric response is still strange. In other words, there is no way to explain why these movements are synchronous. However, that does not impact the possibility of forming a reliable forecasting protocol from the observations of these experiments. The case studies in this book support the phenomenological approach but in no way try to explain, in a cause-and-effect context, the basis for the strange attraction between planetary motion and atmospheric phenomena. It seems, however, that the planetary movements and atmospheric responses shown in this section of this work constitute a primer describing the fundamental constitution of the soul of the Earth.

The following images show climate phenomena that are linked to the movements of the Moon through time. The lunar orbit is a rhythmic phenomenon. The migration of air masses is also a rhythmic phenomenon. The juxtaposition of the movements of the Moon with abrupt shifts in stable air masses makes it possible to track the movements of the Moon by watching the atmosphere over the northern hemisphere.

These charts are from National Oceanographic and Atmospheric Administration (NOAA) sources. They show the atmosphere at the 500mb level over the northern hemisphere. The first set is from September 24, 2000, to September 28, 2000.

The first chart shows the northern hemisphere on Sept. 24, 2000. High pressure is yellow to red and low pressure is blue to violet. Strong highs are red in the center. Weak highs are yellow. Strong lows are violet in the center. Weak lows are turquoise.

On the 24th (second chart on the left), high over Alaska (1), a low over Colorado (2), and a strong low over northern Canada (3) are shown in the image below the first.

Below the second image, we see that, on the 25th, these air masses show simultaneous weakening over the whole Northern Hemisphere.

On the 26th (top right), the weakening of all of the air masses over the hemisphere has progressed even further (top right).

On the 27th (middle right), a sudden shift occurs and all air masses begin to expand rapidly. Note the intensification of the high over Alaska (1), and the low over New England (2).

On the 28th (bottom right), a sudden blooming of air masses has occurred. This phenomenon has been seen over and over in our research. What is happening here? On the 24th, the Moon was at perigee, the closest position in the month to the Earth. That was coincident with the contraction. Two days later a strong expansion follows the contraction. This is an example of the two-day surge.

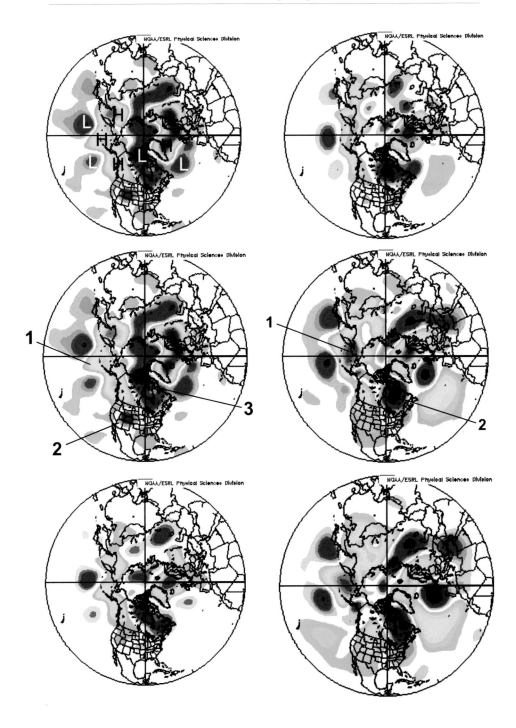

Here is another example of the two-day surge, from April 23 to April 29, 1990. The chart for the 23rd shows stable air masses (right).

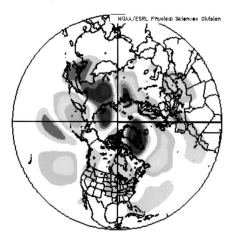

The chart for the 24th is also a stable pattern (top left).

On the 25th, the Moon is at perigee and at the dark on the same day (center left). Note the shift of intensity in the high over the Gulf of Alaska (1) and the low over the Southwest (2). The simultaneous phenomena of perigee and the dark are coincident with a consolidation of the air masses over Alaska, the Great Lakes, and the Southwest.

On the 26th, these masses weaken (bottom left).

On the 27th (top right), two days after the perigee/dark moon event a strong streaming between the PNW and the Southwest (2) and Alaska and the east Pacific (1) can be seen. This is the prelude.

There is a remarkable deepening on the 28th for all air masses in the hemisphere. Note the intensity of the high over Alaska and the low over the Great Plains.

The intensification continues on the 29th (bottom right). This is an example of consolidation. A strong event like the coincidence of perigee and the dark is often accompanied by a consolidation with a strong expansion in two or three days. The two-day surge is not an exact phenomenon of 24 hours, but more of a tendency for the atmosphere to react to lunar rhythmic phenomena as a whole within a certain time frame.

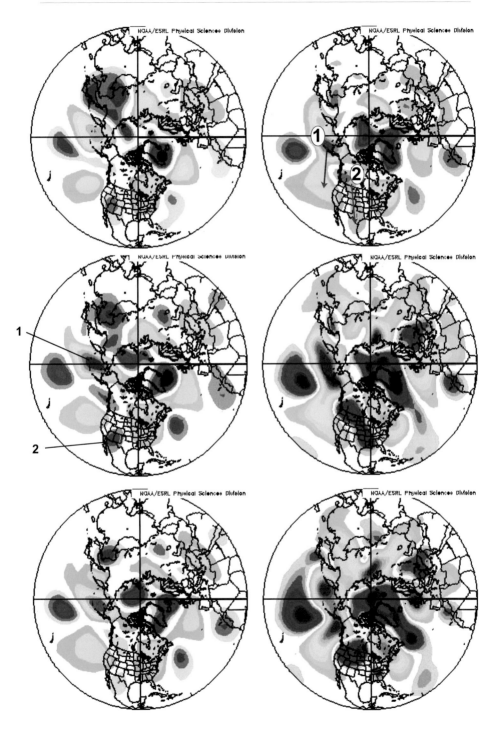

This next series shows a full-moon response on October 12, 2000, unfolding and then moving into a perigee response a week later on October 19.

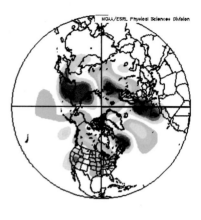

The full moon was late in the day on the 12th (top right).

On the 13th (center right), the contraction of the air masses has already begun.

On the 14th contraction is complete (bottom right).

Strong expansion on the 15th in all areas (opposite top left).

Strong expansion continued on the 16th (opposite second left).

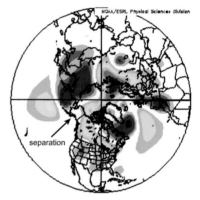

Weakening on the 17th (third left).

Continued weakening on the 18th (bottom left).

On the 19th, the Moon is at perigee (opposite top right). There is another contraction and weakening of air masses.

Weakening after perigee (opposite second right).

Weakening after perigee (opposite third right).

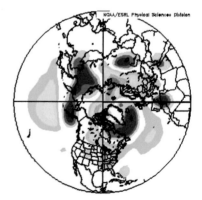

Right on time, the vigorous expansion occurs on the 22nd (opposite bottom right).

Of course there are many more aspects to these phenomena, but these examples illustrate a general linkage between the lunar movements that are coincident with the hemispheric responses of air masses.

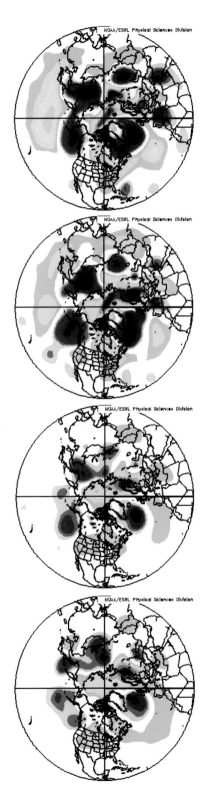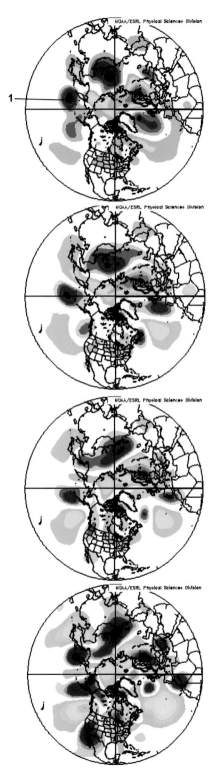

Lunar transits

The next set of charts depicts a lunar movement that is also coincident with an observable atmospheric phenomenon. This pattern links the movements of the Moon in celestial longitude with air-mass movements in particular time frames. It is the phenomenon of lunar transits. The first chart shows the air masses present on December 12, 2004. A strong low is off the coast of California, and another well-developed low is over the Great Lakes and New England. A strong high is over Alaska.

In the chart for the 13th, the low over the Aleutians has suddenly merged eastward into the central Gulf of Alaska, and the high over B.C. has flowed eastward into the Dakotas. The low over the Great Lakes is flowing eastward into the Atlantic.

On the 14th (bottom chart), the low over the Pacific has drawn back to the west and the high over the High Plains is strengthening as the low over New England moves east over the Maritimes.

In the chart for the 15th (opposite top left), the high over the High Plains has moved east into the Central states. The low over New England has moved near Greenland and the high that was in the Anchorage area on the 14th has moved east into the PNW.

By the 16th (opposite top right), the PNW high is moving east toward the Dakotas, the high over the central states

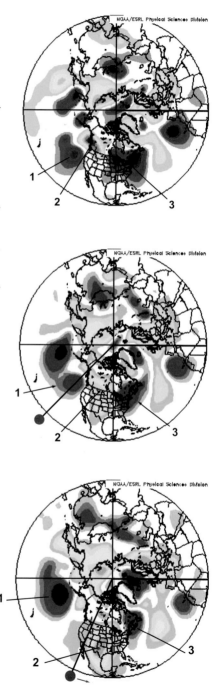

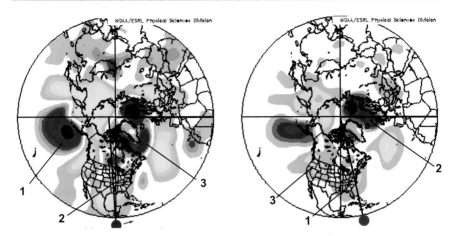

has moved east into the Carolinas, and the low over Greenland has moved east into the north Atlantic. All of these movements since the 13th have been consistently toward the east. What is behind these concerted movements?

The composite below shows the previous movements of just the high-pressure air mass from the west linked to the passage of the moon across the U.S. during that time frame. Over many years, by watching the onset of the eastward flow of all air masses at specific times, it has been recognized that the movement of the moon in particular longitudes is often coincident with the eastward migration of air masses across the U.S. Over time, these periods have become identified as lunar transits.

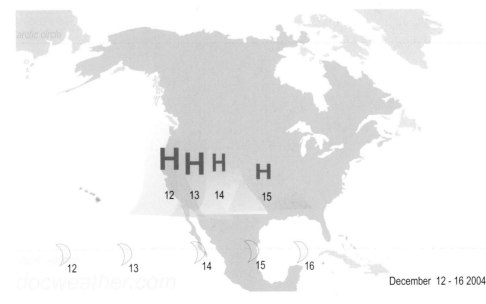

December 12 - 16 2004

Jan. 7 to 13, 2005: This set of charts depicts a variation of the lunar transit. In this variation, the air masses do not move eastward, but each one in the path of the transiting Moon is the site of a sudden intensity that fades as the moon moves to the east.

Jan. 8, 2005: In the first chart, for January 8, 2004 (top left), the position of the moon is depicted by a line with a red circle on the end. The moon in passing the Aleutians is in the longitude of a strong high-pressure surge on that day.

Jan. 9, 2005: In the chart for the 9th (center left), the sweep line of the moon is in the longitude of the eastward bulge of the high over the Aleutians. The high is now streaming to the east to enter western Canada. The Hawaiian tail of the PNW low is accompanying the eastward moving Moon line.

Jan. 10, 2005: On the 10th (bottom left), the tail of the Hawaiian low is gone as the Moon moves against the West Coast. The Aleutian high has receded again to the west, but a circumpolar low northeast of Alaska is following the Moon into northern Canada.

Jan. 11, 2005: On the 11th (top right), the Moon has moved onto the continental U.S. The high-latitude low in northwestern Canada has become a strong element in the chart. The low over the PNW has also come on land and is moving eastward in the wake of the Moon line. The continental high over the eastern half of the nation is receding east in front of the advancing Moon line and bunching up over the eastern third of the country.

Jan. 12, 2005: On the 12th (center right), the Moon line is over the continental divide, and the western third of the country is buried in low pressure, while the northeast high begins to intensify with the approach of the Moon.

Jan. 13, 2005: The Moon has crossed the Mississippi on the 13th (bottom right), and the strong low pressure to the west is fading. The high to the northeast is intensifying with the lunar approach and is showing signs of moving off the coast of the Maritimes as the Moon transits into the Atlantic on the next day. These patterns have been observed numerous times and are often a reliable indicator of potential frontal passages during the month.

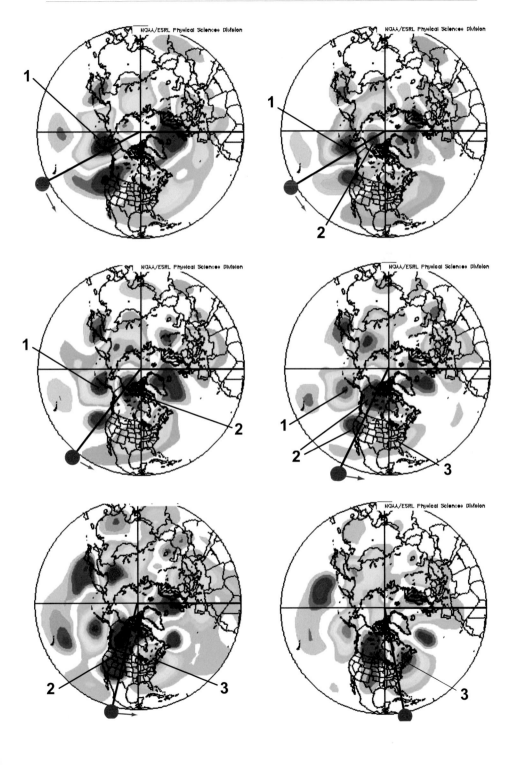

This next set of charts depicts a lunar movement that is also coincident with an observable atmospheric phenomenon. This pattern links the movements of the Moon in celestial longitude with air mass movements in particular time frames. It is the phenomenon of lunar transits.

The first chart shows the air masses present on December 12, 2004. A strong low is off the coast of California and another well-developed low is over the Great Lakes and New England. A strong high is over Alaska.

In the next chart, for the 13th, the low over the Aleutians suddenly has merged eastward into the central Gulf of Alaska, and the high over B.C. has flowed eastward into the Dakotas. The low over the Great Lakes is flowing eastward into the Atlantic.

On the 14th (third chart), the low over the Pacific has drawn back to the west and the high over the High Plains is strengthening as the low over New England moves east over the Maritimes.

In the next chart (opposite top left) for the 15th, the high over the High Plains has moved east into the Central states. The low over New England has moved near Greenland, and the high that was in the Anchorage area on the 14th has moved east into the PNW.

By the 16th (opposite top right), the PNW high is moving east toward the

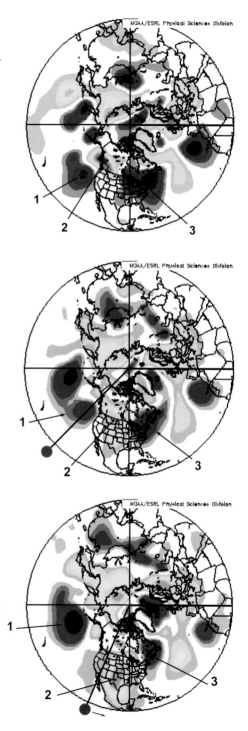

Dakotas, the high over the central states has moved east into the Carolinas, and the low over Greenland has moved east into the north Atlantic. All of these movements since the 13th have been consistently toward the east. What is behind these concerted movements?

Many years by watching the onset of the eastward flow of all air masses at specific times has made it clear that the movement of the Moon in particular longitudes is often coincident with the eastward migration of air masses across the U.S. (bottom image). Over time, these periods have become identified as lunar transits. In this series, the air masses are moving from west to east, following the transit of the Moon across the continent.

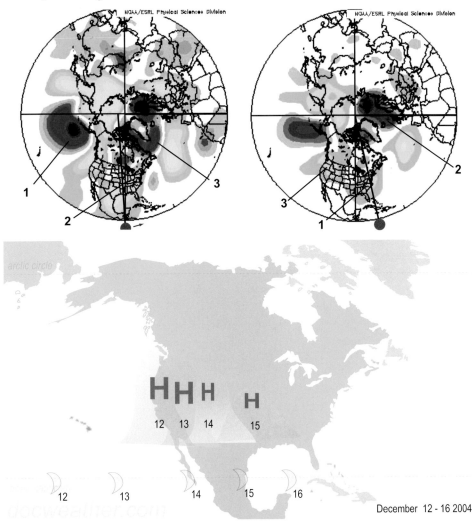

December 12 - 16 2004

This set of charts depicts a variation of the lunar transit. In this variation, the air masses do not move eastward, but each one in the path of the transiting Moon is the site of a sudden intensity that fades as the Moon moves to the east.

In the first chart, for January 8, 2004, the position of the Moon is depicted by a line with a red circle at the end. The Moon in passing the Aleutians is in the longitude of a strong high-pressure surge on that day.

Next, in the chart for the 9th, the sweep line of the Moon is in the longitude of the eastward bulge of the high over the Aleutians. The high is now streaming to the east to enter into western Canada. The Hawaiian tail of the PNW low is accompanying the eastward moving Moon line.

On the 10th, the tail of the Hawaiian low is gone as the Moon moves against the West Coast (bottom chart). The Aleutian high has receded again to the west, but a circumpolar low northeast of Alaska is following the Moon into northern Canada.

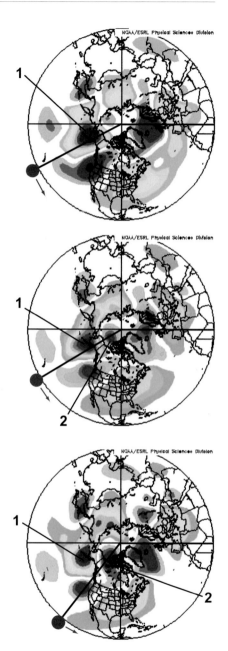

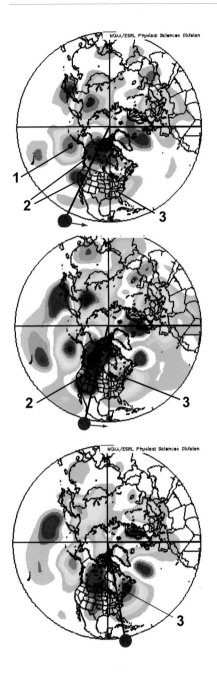

On the 11th (top chart), the Moon has moved over the continental U.S. The high-latitude low in northwestern Canada has become a strong element in the chart. The low over the PNW has also come on land and is moving eastward in the wake of the Moon line. The continental high over the eastern half of the nation is receding east in front of the advancing Moon line and bunching up over the eastern third of the country.

On the 12th (center left), the Moon line is over the continental divide and the western third of the country is buried in low pressure, while the northeast high begins to intensify with the approach of the moon.

The Moon has crossed the Mississippi on the 13th, and the strong low pressure to the west is fading (final chart). The high to the northeast is intensifying with the lunar approach and is showing signs of moving off the coast of the Maritimes as the Moon transits into the Atlantic on the next day. These patterns have been observed numerous times and are often a reliable indicator of potential frontal passages during the month.

Moon line / Sun line

Several decades of observing these phenomena have gradually made it possible to form charts that can accurately locate planetary positions on Earth, even for slower-moving planets like the Sun. The following charts show the work of Jonathan Code, a researcher in the U.K. He had been researching high-latitude frontal passages and discovered a pressurizing pattern linking the positions of the transiting Moon with the more sedate position of the Sun.

In the first chart, for December 15, 1998, the Sun (red line) is over the eastern Pacific, while the Moon (blue line) is over the central Pacific.

Next, on the 16th, the Moon has moved into the longitude of the Bering Sea. The small low south of the Aleutian chain is showing the signs of the lunar transit. Most interesting here is that the high-pressure area on the West Coast near the Sun line is showing signs of becoming more intense.

On the 17th (bottom), the Moon is near Hawaii, and the low to the northwest of the islands is showing signs of the transit. The

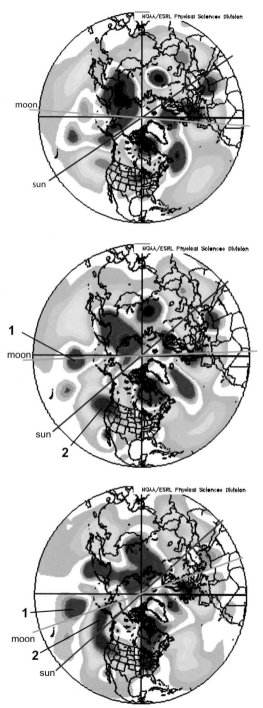

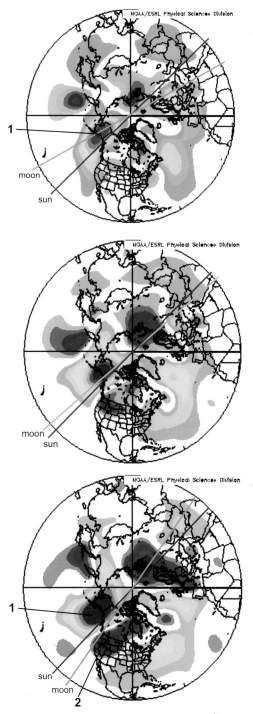

ridge against the coast is being squeezed against the Sun line and shows a remarkable intensification.

On the 18th (top left), the Moon has passed Hawaii. The high-pressure area over Alaska is now squeezed between the Sun line and the Moon line.

On the 19th (center), the Moon is just about to cross the Sun line and the high-pressure area is modeling itself almost exactly to the area where the two lines are about to cross.

On the 20th (bottom), the Moon has crossed the Sun line and the ridge over Alaska is very intense. The low to the east of the Sun line suddenly deepens as the Moon transits its longitude. In this pattern, the air masses do not migrate with the motions of the Moon, but rather the Sun line acts as a foil to their eastward progress. The significance of these observations is that it is possible to locate a particular planetary influence on the atmosphere with a particular location on Earth. This observation made with the Sun and Moon will be applied to Mercury in the next section.

Mercury retrograde motion and the jet stream

Nov. 18 to Dec. 3, 2002: This chart is a depiction of the intensity and direction of the winds in the upper levels of the atmosphere. The section is at the 500mb level, about halfway up. The winds that guide storms as they travel along the Earth surface are found at that level. The chart shows the upper air wind currents for the time period between November 18 and December 3 of 2002. A very strong current (red) is flowing into Alaska from the south. Another, moderately strong current (yellow) is in the central Pacific. Using our technique for placing a planet into a specific terrestrial longitude, we can place the planet Mercury just to the west of Hawaii during this time period. That would put it in the center of the moderate current in the mid-Pacific.

Dec. 3 to Dec. 17, 2002: The second chart shows the vector wind currents for the time period of December 3 to December 17. Mercury moved into the eastern Pacific at that time, and we can see that the center of the moderate current has moved along with it into a position just off the West Coast.

Dec. 17 to Dec. 31, 2002: The next chart shows the time period between

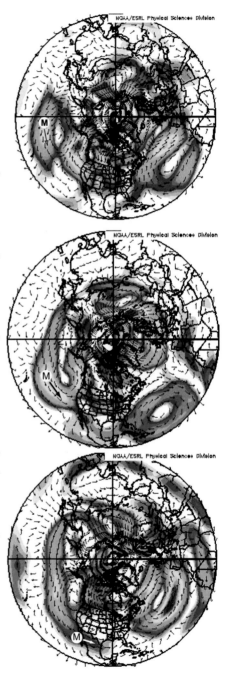

December 17 and December 31. Mercury crossed into Central America at that time, and the center of the current falls back to Hawaii, while the focus of the current is to the south, into Central America.

Jan. 1 to Jan. 23, 2002: In the next chart (below left), we see a rather startling image. The immense vortex near Hawaii has very strong winds. It is as if something has suddenly blocked the jet stream from flowing to the continent from the ocean. The formation of this remarkable vortex was exactly coincident with a retrograde motion of Mercury over the western U.S. during this time frame. Retrograde motion is the apparent backward movement of a planet as seen against the background of the fixed stars.

Jan. 23 to Feb. 3, 2002: This was the time frame when Mercury once again began to move in a west-to-east direct motion. The great eddy near Hawaii has dispersed, and a strong easterly wind accompanies the motion of Mercury across the southern Rocky Mountains. Retrograde motion has proved to be a significant element for the configuration of upper-level winds. The position of a retrograde planet in a particular longitude often has a strong impact on seasonal anomalies such as hurricanes, to be discussed later.

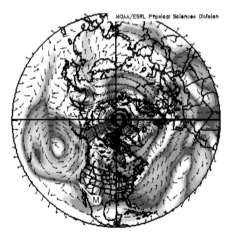
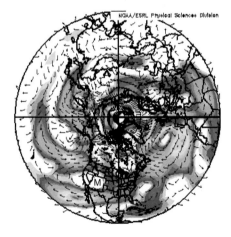

El Niño-related Pacific Phytoplankton Bloom (center), July 1998
(NASA/Goddard Space Flight Center, The SeaWiFS Project and ORBIMAGE)

CHAPTER 3
A NEW LOOK AT EL NIÑO

To the casual observer, it may come as a surprise to learn that the phenomenon known as "El Niño" is actually a yearly event. In fact, a mass of warm water builds each winter in the western Pacific. During spring every year, this warm water moves eastward into the region of the dateline at longitude 180°

This annual buildup (as shown in the image) and subsequent migration is a regular phenomenon in the western Pacific. Occasionally the buildup and migration of the warm water into the central Pacific finds conditions that support the further eastward migration of the warmth pool. This happens when conditions at the dateline and in the eastern Pacific support the continued eastward migration of the warm water from the central Pacific. It is known that the conditions that support enhanced eastward flow are rhythmic in decadal time frames. This series of movements is known to climatologists as the canonical El Niño. However, the precise timing of the enhanced flows is not understood. If no support for enhanced eastward flux occurs at the dateline in midsummer, then El Niño will not move into the west coast of South America. No clear phenomenon has presented itself as the determining parameter of these patterns. In the following charts the archetypal movements of the warm water during a non-El Niño year is illustrated.

The basic chart (top of page 84) divides the Pacific into five different areas. El Niño 1+2 is the area just off the west coast of Peru. Niño 3 is the area in the eastern Pacific to the east of Hawaii. Niño 3.4 runs from just east of Hawaii

date														

110	120	130	140	150	160	170	180	170	160	150	140	130	120	110	100	90	80

niño 5	niño 4	niño 3.4	niño 3	1+2
leo	virgo	libra	scorpio	sagittarius	capricorn	

westward to the dateline at 180° of longitude, west of Hawaii. Niño 4 runs from the dateline westward to the east coast of Australia. What we will call Niño 5 runs from the east coast of Australia westward to Indonesia.

3 January, February, March														

warm

110	120	130	140	150	160	170	180	170	160	150	140	130	120	110	100	90	80

niño 5	niño 4	niño 3.4	niño 3	1+2
leo	virgo	libra	scorpio	sagittarius	capricorn	

In January, February, and March warmth builds in Niño 5.

4 April														

warm

110	120	130	140	150	160	170	180	170	160	150	140	130	120	110	100	90	80

niño 5	niño 4	niño 3.4	niño 3	1+2
leo	virgo	libra	scorpio	sagittarius	capricorn	

In April the warm water moves into Niño 4 and to the edge of Niño 3.4, near the international dateline.

In May, cooling emerges in Niño 5, and the warmth shifts eastward to the middle of Niño 3.4.

In June, the cooling settles into western Niño 4, and the warmth settles into eastern Niño 4 and central Niño 3.4.

In July, the cool expands in Niño 4, and the warmth shifts to the east and begins to shrink.

8 August, September

neutral

110	120	130	140	150	160	170	180	170	160	150	140	130	120	110	100	90	8 0

niño 5	niño 4	niño 3.4	niño 3	1+2

leo	virgo	libra	scorpio	sagittarius	capricorn	

In August and September, the cool and warm masses approach each other at the dateline.

9 October to December

neutral

110	120	130	140	150	160	170	180	170	160	150	140	130	120	110	100	90	8 0

niño 5	niño 4	niño 3.4	niño 3	1+2

leo	virgo	libra	scorpio	sagittarius	capricorn	

In October and November, the cool water to the west meets the warm water to the east at the dateline.

10 December

cool   warm

110	120	130	140	150	160	170	180	170	160	150	140	130	120	110	100	90	8 0

niño 5	niño 4	niño 3.4	niño 3	1+2

leo	virgo	libra	scorpio	sagittarius	capricorn	

In December, the cooling and warmth neutralize each other.

Occasionally, in a little-understood periodicity, the warmth at the dateline does not neutralize but continues moving east, eventually ending up on the west coast of South America. Many physical hypotheses have been presented to explain this mysterious periodicity. Research has shown that increased convection from numerous clustered thunderstorms at the dateline during spring supports the eastward continuation of the warmth, sending subsurface Kelvin waves into the east Pacific. The physical cause for the increase in convection at the dateline is clearly that there is more warmth in the water to support convection. The question still unanswered by the physical data is: Why is so much warmth available at the dateline in some years and not in others?

In the following sequence, we will see the winter buildup and spring dateline migration described in the preceding sequence. The difference is that, instead of fading at the dateline in the spring, the warmth will be enhanced and migrate toward Hawaii as spring unfolds into summer. Furthermore, this movement continues into the southeast Pacific, bringing warm water to this region by Christmas. This is the El Niño that most people recognize.

This sequence is reflected in the first few images of this chapter. From January to June, warm water builds first in Niño 5 and 4. Then, by April, Niño 4 warms as Niño 5 begins to cool. In May, the cold water spreads eastward into Indonesia as warm water spreads eastward into Niño 4. In June, the cool grows in Niño 5, as the warmth migrates into the whole of Niño 4 and approaches the dateline. So far the patterns are the same.

During El Niño years, there is a shift at this time to a new pattern. This shift distinguishes a canonical El Niño from an El Niño event. In an El Niño or warming event, the warmth at the dateline moves east of the dateline for the rest of the year instead of neutralizing at the dateline in July.

In an El Niño July, the cold water spreads into Niño 4, bringing drought to the northern region of Australia. Warmth spreads toward Hawaii through Niño 3.4. By August, a cold pool spreads through the western Pacific, while warmth travels east.

From September to November, cold settles into place in the west as the warmth expands and spreads into the eastern Pacific. By December, the entire Pacific, from the dateline to Peru, is the site of a large, warm pool of water, and El Niño reveals its impact on world weather.

11 El Niño / August

cool				warm					

110	120	130	140	150	160	170	180	170	160	150	140	130	120	110	100	90	8 0
niño 5		niño 4					niño 3.4						niño 3			1+2	
leo		virgo		libra		scorpio		sagittarius		capricorn							

12 El Niño / September to December

cool				warm					

110	120	130	140	150	160	170	180	170	160	150	140	130	120	110	100	90	8 0
niño 5		niño 4					niño 3.4						niño 3			1+2	
leo		virgo		libra		scorpio		sagittarius		capricorn							

This sequence is the El Niño warmth event. Research has revealed this pattern, and many indicators of it, but attempts to find reliable related periodicities in other phenomena have proved to be statistically weak. The complexity of the possible physical elements in such a vast time and space system is staggering. We could ask: Is there a system in which the natural periodicities resemble the periods in the El Niño cycle?

Contemporary research in climatology is focused on finding such periods in the physical interaction of currents or in atmospheric/sea-surface temperature linkages. These researches focus on ever smaller micro-data inputs, hoping for a symptom at the micro level that can explain the oscillations at the macro level. This approach multiplies the variables within the system, which greatly enhances the possibility for error. Perhaps a research strategy that would yield some useful insights would involve looking for macro periods even longer than the time periods in which the phenomena unfold. Since the periodicities of El Niño and La Niña phenomena are quasi-biennial, or even inter-decadal, it would seem reasonable to look for large-scale periodicities with these time signatures.

The greatest and most predictable source of large scale time signatures available is the system of movements found in the orbits of the planets. It seems reasonable that, if a natural phenomenon being studied manifests in periods of ten or twelve years, a juxtaposition of these events with others occurring in periods of ten or twelve years would yield insights that could be statistically significant. This is exactly what was done in this study.

The El Niño warmth event and its periods were placed into a context of the direct and apparent retrograde motion of planets transiting the Pacific in a given year. This approach helped to reveal a strong correlation between the direction and timing of a planet in a given sector of the Pacific and the onset of El Niño. The initial theoretical insight was then applied to actual El Niño events and La Niña episodes in a continuous study from 1976 to 1998. A significant degree of correlation was found between the periods of planetary motion in a given sector and the particular climatic response.

We can recall that in the early part of the year in the canonical El Niño there is a strong buildup of warmth in the sea water in the Niño 5 sector of the Pacific Ocean. Physical research by satellite reveals that in this area a vast hillock of water actually mounds a meter or two higher than the sea level throughout the rest of the Pacific. Some force appears to be pushing from east to west and mounding the water. Physical influences such as the action of the easterly trade winds (east to west) in the Niño 4 sector have proved to be unreliable predictors of the onset of this phenomenon. This is especially so because, in fact, the winter/spring buildup in Niño 5 occurs in many years other than in El Niño years, and it occurs independently of the quasi-biennial trade-wind oscillation.

Looking at the winter/early spring buildup from a perspective of planetary motion, we can formulate protocols for prediction that have been determined through observation. The first is that a planetary position in celestial longitude can be projected onto the Earth. These projected positions are often coincident to extreme weather events. As a result of this projection technique, it is possible to correlate positions in celestial longitude with weather events in terrestrial longitude. This projection method finds further validation when a planet at a particular position in celestial longitude moves into retrograde motion. A blocking high often forms in the projected longitude of the retro-grade loop. This frequently observed phenomenon was the original stimulus to form an El Niño model based on retrograde motion in specific longitudes at specific times.

Regarding El Niño, it can be observed that any outer planet in the longi-tude of Niño 5 in the western Pacific will have a retrograde period that is coin-cident with the winter and early spring warmth buildup in that area. In the next image, this retrograde motion is illustrated using Jupiter as an example.

Retrograde motion is depicted with an arrow pointing to the left, or west, of Jupiter, accompanied by "RX." Direct motion is represented by an arrow to the right, or east. The arrow is accompanied by "D." The period of an outer planet in Niño 5 would be retrograde in January and direct in May. As a result, the retrograde motion (i.e., east-to-west motion) of any outer planet in Niño 5 is coincident with the east-to-west movements of warm water in that far-western sector each year. This was depicted in the second figure labeled January, February, March earlier in this chapter. Linking this canoni-cal motion of the warmth plume with the movement of the outer planets in a given year we see that the onset of west-to-east direct motion of any outer planet in that sector is coincident with the onset of the west-to-east migration of warm water out of Niño 5 in the canonical year. This pattern, in which outer planets are active in Niño 5, is present in most significant El Niño events.

The next figure depicts the retrograde and direct periods of any outer planet from March to July near the dateline. The dateline falls on the border between Niño 4 and Niño 3.4. This retrograde motion coincides with the crit-ical dateline support needed for the development of strong El Niño patterns.

14 March retrograde motion

8 D →
Jupiter
← RX 3

110 120	130 140	150 160	170 180	170 160 150	140 130	120 110 100	90	80
niño 5	niño 4			niño 3.4		niño 3	1+2	
leo	virgo		libra	scorpio	sagittarius	capricorn		

In the eastern Pacific, the retrograde and direct motion of any planet in Niño 1+2 coincides directly with the canonical El Niño as planets in the far-eastern Pacific go retrograde in June, building up warm water to the west (the tropical Pacific east of Hawaii). Outer planets in Niño 1+2 then go direct in either December or January, coincident with a west-to-east flow that marks the onset of warming sea surface temperatures in Niño 1+2.

15 December retrograde motion

11 D →
Jupiter
← RX 6

110 120	130 140	150 160	170 180	170 160 150	140 130	120 110 100	90	80
niño 5	niño 4			niño 3.4		niño 3	1+2	
leo	virgo		libra	scorpio	sagittarius	capricorn		

The next chart shows a composite of the retrograde and direct rhythms of the planets in a month-to-month pattern. This rhythm will then be overlaid with Sea Surface Temperature (SST) data to illustrate the high degree of coincidence in the two systems. In January of each year, any outer planet placed in the eastern half of Niño 5 would have a retrograde period beginning in January and ending the following June. This period of retrograde and direct motion holds true for any outer planet in Niño 5 in any given year. The inner planets (Mars, Mercury, and Venus) move much more rapidly through their orbits and, as a result, constitute more rapid and local movement phenomena than do the outer planets.

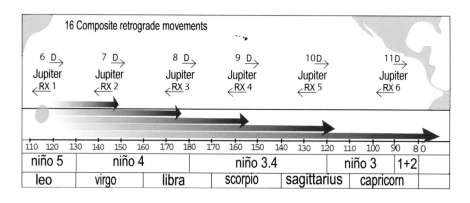

From the composite we can see that any outer planet in central Niño 4 near the dateline will go retrograde in February and direct in July. Any outer planet in the western half of Niño 3.4 will go retrograde in March and direct in August. Any outer planet in the eastern half of Niño 3.4 will go retrograde in April and direct in September. Any outer planet in Niño 3 will go retrograde in May and direct in October, and any outer planet in Niño 1+2 will go retrograde in June and direct in November or December.

These patterns of retrograde and direct motion have proved to be highly coincident with the seasonal fluctuations of warm and cool water in the Pacific. This was studied by gridding month by month SST values in all sectors of the Pacific since 1980. Retrograde and direct-motion values and time frames for each month were then integrated into the SST values. Strong correspondences in this study provided the basis for this article.

Since the pool of warmth builds in the western Pacific every year, and every year there is a gathering of warmth at the dateline in the spring, it seems logical that there is an influence that supports this. Some years, the pool of warm water at mid-Pacific finds support for flowing to the east at midsummer in the end of June. This is precisely the time period for a shift in direction of any outer planet positioned in celestial longitude near the dateline.

Many case studies of these phenomena have been made. A short animation was made of the 1982/1983 and 1997/1998 El Niño events and of several La Niña occurrences. These studies have shown consistently that the placement of planets over the Pacific and the timing of their retrograde and direct motions has a strong coincidental correspondence to the movements of the Sea Surface Temperatures in the longitude of the movements.

This is the general rule. However, this pattern is affected by two variables. The first is that initial conditions greatly effect whether it is warmth or cold that is in flux. If cool water is to the west when a planet goes direct, then the records show that there will often be cool water going to the east for the next few months. If warm water is to the west when a planet goes direct, then, the records show that there will often be a coincidence of warm water going to the east in the next few months.

The second variable in which a disturbance of the flow of water is often coincident with a shift in planetary motion is the intrusion of a loop of Mars, Mercury, or Venus over the Pacific in a given year. These movements greatly modify the canonical El Niño into biennial and decadal rhythms. The record El Niño of 1997/1998 was stopped dead in its tracks by a combined retrograde motion of Mercury and Venus in the eastern Pacific in December of 1997. Only when they went direct did the weather patterns influenced by El Niño move into the coast. The looping rhythms of the inner planets have proved to be coincident with such unusual conditions in a given season. The qualities of the effects are the same for inner planets as they are for outer planets. The effects are just placed into slower or faster time frames.

The classic El Niño events of 1982/1983 and 1997/1998 have been the subject of a much more detailed study presented in another section of this chapter. Those years, and the unusual Pacific warming trend of the 1990s, fit very easily into the parameters given in this survey. When working with this model, remarkable coincidences between planetary motion and the shifts of the ocean–atmosphere linkages can often be observed.

MORE ON EL NIÑO

In this section, we will use the term *surge* to describe the occasional motion of an anomalous west-to-east warmth-pool migration from the dateline to South America during the fall of the year. This surge can be modeled in various ways, depending on the placement of other planets across the Pacific. If the surge is to be supported, there are two variations of motions that can be observed. The first is the RX , or retrograde, block.

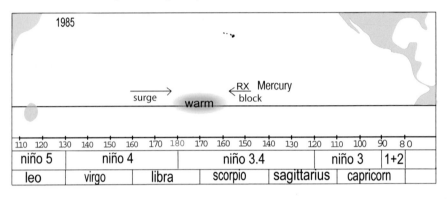

The RX block is a pattern that occurs when the surge is met by the retrograde motion of another planet that intercepts the surge, stops its eastward momentum, and amplifies it. The chart above is from 1985. The RX block of Mercury in Niño 3.4 in December formed a warmth spike in the mid-Pacific in an otherwise cold year. We can see an image of this in the motion of water in a bathtub. If you move your hand from one end of the tub to the other in a slow rhythmic way, the water eventually begins to surge rhythmically back and forth, from one end to the other, in an oscillatory pattern. The wave period is related harmonically to the dimensions of the bathtub. The surge of a rhythmic wave is an image we can use to describe the surge of an El Niño during a canonical year. If you put your hand in the tub and hold it in one place, your hand, as a stationary force, resists the surge of the oscillating waves of water. This is a good image for the observed blocking pattern that normally sets up in the atmosphere when planets are on station (apparently motionless) in a given longitude.

When such a surging oscillatory wave is started and then, as it pulsates from left to right (direct motion of a planet), you put your hand into the

1985 block / spike

surge → warm ←RX Mercury
block cool

110	120	130	140	150	160	170	180	170	160	150	140	130	120	110	100	90	8 0
niño 5		niño 4					niño 3.4					niño 3			1+2		
leo		virgo		libra		scorpio			sagittarius			capricorn					

middle of the tub and begin to push from right to left (retrograde motion of a planet), you are exhibiting a counter-force to the motion of the surge. The resistance to the surging flow forms strong turbulence. Using this analogy in larger bodies of water, we can observe that, when surges form in the top layer and are met with a counter-force, the interacting force fields of the two movements manifest as turbulent peaks. In hydrology this pattern is called *chop*.

In the ocean, when the canonical El Niño is rhythmically pulsing through the upper layers of the Pacific basin, there is a yearly arising of the warmth in the west in the period from January to March. Then, there is usually a slow west-to-east surge of the warmth to the dateline in July/August. Sometimes the warmth surge moves across the whole Pacific and ends up on the west coast of South America, setting up an El Niño. If a planet goes into retrograde motion somewhere between the warmth surge at the dateline and the coast of South America, a consistent tendency has been observed for the SSTs in the longitude of the retrograde planet to rise. The rise in SSTs in the longitude of a retrograde planet is known in this system as a *warmth spike,* or "squeeze." When a number of well-placed RX blocks occur in a given year, there is great support in that year or in the next for the occurrence of El Niño events. The compression, or squeeze, of the chop is often accompanied by spikes in SSTs that remain as warmth pools for subsequent development.

A more potent RX block pattern can arise when a transiting planet that has just gone into direct motion in a western position accompanies a surge from west to east. Then, the new forces of the direct motion amplify the surge momentum. This becomes especially potent when the amplified west-to-east surge is met with an RX (east-to-west) block from a planet to the east.

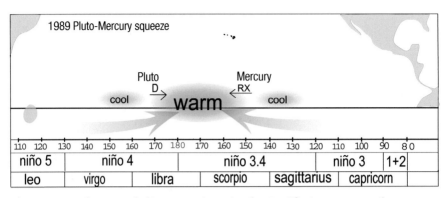

The timing of RX and direct motions in the Pacific is prone to these types of events, since the annual motion of retrograde-to-direct motion proceeds across the Pacific in a west-to-east direction. A typical squeeze can be seen in a chart from 1989.

The motion of a planet moving from retrograde (east-to-west) motion into direct (west-to-east) motion is often accompanied by observable changes in SSTs. When the timing and positions of the retrograde and direct movements of two planets put them close enough in longitude for the recently direct motion of the western planet to encounter the retrograde motion of a planet to the east, then the warmth spike is highly amplified into a squeeze. A squeeze involves strong spikes in SSTs in the longitude between the two interacting planets. In 1989, Pluto went direct (west-to-east) in late July at the dateline, and Mercury went retrograde (east-to-west) in western Niño 3.4 in early September, as Pluto was still moving from west to east. The result was a squeeze accompanied by a strong spike in SSTs in Niño 3.4 at that time. When a number of such strong spikes in SSTs occur during a particular year, it often enhances the warmth of a surge that is underway. This was the case late in 1989.

Accompanying a squeeze (see next image) is another phenomenon that is also problematic in the onset of El Niño–La Niña conditions. When the squeeze is underway, the longitude to the west of the western planet often is the site of falling SSTs. It is as though the compression forces of the squeeze focalize the warmth between the two squeezing planets at the expense of the areas adjacent to the squeeze. If too many squeezes generate warmth in one year, then it appears as if a "warmth–cold scar" forms in the longitude of the

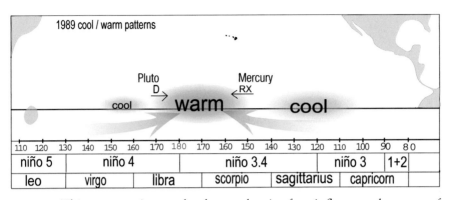

squeeze. This scar persists to the degree that it often influences the onset of El Niño or La Niña conditions the following year. The placement of squeezes and RX blocks across the Pacific and their timing in the context of the canonical El Niño present a rich array of possibilities for modeling the rhythmic oscillation of El Niño and La Niña events. This type of pattern is especially evident in the development of a classic El Niño. In the chart, a warmth pool near the dateline persisted into 1990, serving as a warmth seed at the dateline for the unfolding of El Niño in the summer of that year.

In El Niño, the planets across the Pacific were positioned in such a way that their direct and retrograde motion allowed the warmth surge to build up in the west early in the year and then, as the western-most planet (Saturn) went direct, the warmth surge moved eastward just in time to meet the next planet (Jupiter) to the east, just as it turned into direct motion. In all, five major planets were placed perfectly to support the west-to-east motion of warm water across the Pacific. This pattern was further supported by the rapid transit of Mars across these movements between May and December of that year. The result was a remarkable warming of the eastern Pacific during the end of 1982 to 1983.

In the great El Niño of 1997 to 1998, it was also the case that Mars made a dynamic transit across the Pacific in the months leading up to the event. Between 1983 and 1998, the cluster of planets in the mid-Pacific had moved into the eastern Pacific. Owing to squeezing rhythms in the previous years, the summer of 1997 was a time of great heat development in the central Pacific. In the fall of 1997, Mars, just coming off a looping pattern in the beginning of the year in Niño 4, once again accompanied a strong warmth migration across the Pacific from the dateline in the fall. The cluster of Jupiter, Uranus, Neptune, and Pluto between Hawaii and the West Coast supported a very strong transport of warmth from west to east late in the year.

As an aside, in 1999 this cluster was broken up by the motion of Jupiter into the Atlantic, and by Uranus and Neptune out of the eastern Pacific and into the Gulf of Mexico. Those movements have been coincident with a failure of El Niño to gain strong impetus since then.

In general, the year after a strong El Niño, the Pacific cools rapidly and, unless there are more squeezes to push up the SSTs in the proper rhythm, the cooling dominates. It appears that this is often the case when the Pacific transit of Mars is not happening. Every second year, Mars is not present in the Pacific, owing to it's orbital periodicity. During the non-Mars years, the Pacific surge does not find support in the Mars east–west transit during the winter. This is often a La Niña signal. If a non-transit period of Mars is also coincident with a lack of planets across the Pacific, or with the lack of planets at the dateline in midsummer, or by the awkward placement of RX blocks that dissipate the surge from the dateline, then cold conditions will dominate.

The cooling pattern that results from such blocking arising in La Niña years can often be coincident (as it was from 1983 to 1986) to times when both Mars and Mercury have strategic retrograde periods in the Pacific. In 1984, for instance, Mars had a retrograde motion sequence at the dateline from April to June. This movement interacted with Pluto and Saturn in that area, creating a block that would effectively retard any eastward motion at the dateline during this critical period. It has often been observed that a shift to direct motion of a planet at the dateline in June and July is coincident with the enhancement of El Niño patterns. A block at this time from Mars moving retrograde would tend to neutralize any eastward impetus in June and July.

To create an obstacle to the eastward migration of warm water in that year, Mars showed a promising motion across the Pacific from July onward. The block at the dateline kept SSTs near Hawaii a bit below normal. The slightly cool SSTs migrating eastward toward Hawaii encountered a trio of Uranus, Neptune, and Jupiter east of Hawaii. Mars moved out of the Pacific in November. Often, the exit of a planet across Central America is coincident with a falling of SSTs in the eastern Pacific. Then, early in December, Mercury went retrograde in the midst of the cluster of Uranus, Neptune, and Jupiter in the east Pacific. This was another blocking event in an already-cool eastern Pacific. These blocking patterns were coincident with a moderate mid-Pacific cooling trend in that summer and fall, which set the stage for strong cooling the next year.

1989 La Niña															

RX Venus

RX Mercury

RX Mercury

					cool	Uranus D	Neptune D		cool			

Pluto D

Saturn D

110	120	130	140	150	160	170	180	170	160	150	140	130	120	110	100	90	80

niño 5	niño 4		niño 3.4		niño 3	1+2
leo	virgo	libra	scorpio	sagittarius	capricorn	

This kind of pattern arose again in 1989 to 1990. During the fall of 1989, a retrograde block of Mercury and Venus on an already-cool Pacific interfered with any potential for warmth to arise, and very cool periods ensued. This happened when there were three planets in the eastern Pacific. Uranus, Saturn, and Neptune formed a tight cluster close to the coast in the summer and fall of 1989. They were retrograde from May to September of that year. That in itself could also be seen as the possibility of late warmth making it to the coast. What would be needed is some kind of dynamic push from the west during the July-to-September gestation period for the typical El Niño. Just as these planets went direct in the eastern Pacific in September, Mercury went into retrograde motion over Pluto in the west, creating a squeeze block. This was coincident with a halt to plunging temperatures in that region, but it was not enough to stimulate strong warming to the east. Mars was not in the picture in 1989, but in the Asian portion of its two-year orbit. SSTs in the eastern Pacific began to rise as the Uranus–Saturn–Neptune trio reached its station in September 1989. However, they once again leveled off during a combined Venus and Mercury retrograde block in December and early January in Niño 3. This block leveled the burgeoning warmth surge that was then present in the east Pacific.

It should be emphasized that, in this system, the forces that create La Niña conditions are not essentially different from the forces that create El Niño. The La Niña pattern still has retrograde motion as its fundamental principle; but in La Niña events, the motion is not syncopated to the rhythmic requirements of the canonical El Niño. In La Niña, the RX blocks and squeezes create a rhythmical damping or hindrance to the orderly procession of the warm-water surge from west to east throughout the year. When they are out of synch to the canon, the orderly procession is interrupted. The energy of the surge is dissipated in the form of thermal "scars." At these times, the warmth is dissipated or localized in a particular longitude and cannot build a flow toward the east. Most often, it is retrograde looping patterns for Venus, Mercury, and Mars that creates the scars.

It often happens that simultaneous blocking patterns in the Pacific over several years create conditions for the arising of warmth pools in several areas of the Pacific, though the pools are isolated from each other. This kind of pattern usually results in a neutral-to-cool spread of temperatures at the equator from the dateline to South America. Then, when the motion of the blocking planets becomes more harmonious to the rhythms of the canonical El Niño, the warmth pools coalesce into a full-blown warmth event. In general, years when blocking east of the dateline is strong from July to November, there is little chance for a large warmth event in the eastern Pacific unless a momentous surge is already in process, as in 1997 to 1998. In that case, two blocking events in the eastern Pacific could not stop the surge, but only delayed it until Venus went direct, and then the El Niño was delayed until late January.

El Niño, 1982 to 1983

This section illustrates a pattern of blocking that was coincident with the buildup of warmth in the Pacific, resulting in the 1982/1983 El Niño. The "Z" shaped squiggle under the planet name is a symbol for a blocking event. The warmth event of 1982/1983 emerged in a time frame starting in October 1981 with a Mercury loop in Niño 4. This was coincident with a gradual warming of that area in the next few months. It was in February 1982 however that a strong set of events took place setting the stage for an El Niño. Mercury and Venus had been traveling in tandem toward the West Coast late in 1981. A gradual warming in Niño 1+2 and Niño 3 was taking place. In January, the two planets crossed into Niño 3. It is often the start of a cooling trend when a fast-moving planet crosses into Niño 3. This crossing was the start of a cooling trend. Then suddenly, both planets went into a looping pattern in Niño 3 in February 1982. This was a strong blocking pattern. SSTs in Niño 1+2 to the east of the blocks took a nosedive, but SSTs in Niño 3 to the west of the blocks turned on a dime, started rising, and never looked back. At the same time on the other end of the Pacific a Mars loop in Niño 4 in February 1982 started a warming pattern there that was coincident with the development of a warmth pool in the western Pacific for the rest of the year. We pick up the story from February 1982.

After the blocking pattern finished in the middle of February, Mercury and Venus went direct (eastward) in Niño 3 and rapidly moved out of the Pacific. These movements were coincident with cooling in Niño 1+2, with above-average temperatures in Niño 3 remaining as a remnant of the strong block. In the western Pacific, Jupiter went retrograde (apparent westward) in Niño 4 in the middle of February. Jupiter joined a group of planets composed of Pluto, Saturn, and Mars, now all moving west in the early winter in the western Pacific. This was coincident with a small cool pool at the dateline to the east and an increase of warmth in Niño 4 and beyond into the far western Pacific. By the end of February, these conditions had worked out to a growing neutral condition at the equator itself, with pockets of warmth spread all over the Pacific, most notably in the far west and a weak pool of warmth in the far-eastern Pacific.

Mar. 82

Saturn RX Pluto RX
warm
Mars RX Jupiter RX   Uranus RX Neptune RX

warm   cool   cool

110	120	130	140	150	160	170	180	170	160	150	140	130	120	110	100	90	80
niño 5			niño 4					niño 3.4						niño 3		1+2	
leo			virgo			libra			scorpio			sagittarius			capricorn		

In March of 1982, Uranus and Neptune went retrograde in Niño 3.4. This meant that all planets over the Pacific were moving westward, from Hawaii to Indonesia. Residual warmth from the events of 1981 was present as pockets of warmth in the western Pacific at that time. With the motion of Uranus and Neptune moving west, the warmth in the eastern Pacific began to dissolve and move toward the dateline area. In its place in the eastern Pacific, a cold tongue began to form off Peru and move west along the equator.

This surge of the coastal upwelling was the seed of a cooling event forming along the Peruvian coast. Since all other planets on the chart were moving to the west, the placement of a growing warmth pool near the dateline and to the west, as well as a sharp cool sector in the east, was in line with the model. By the end of March, it looked as if a La Niña cooling event might even be shaping up along the equator in the eastern Pacific.

May 82

D
Mars
Pluto RX
Saturn RX Jupiter RX   Uranus RX Neptune RX

warm warm warm

110	120	130	140	150	160	170	180	170	160	150	140	130	120	110	100	90	80
niño 5			niño 4					niño 3.4						niño 3		1+2	
leo			virgo			libra			scorpio			sagittarius			capricorn		

In April, however, the dateline area began to heat up as warmth from the western Pacific, which had been present during the winter, began to drift canonically to the east and to accumulate at mid-Pacific. April is a big month for the western Pacific. It is normally the time when the shift begins to happen, in which warm water starts to migrate toward the dateline for a June rendez-vous. No changes in motion took place for any planet in April, so the patterns in place during late winter just lingered. However, all four planets were losing their retrograde impetus and slowing for their station periods in May. The Pacific appeared to be holding its breath.

The surge of cold along the equator in the eastern Pacific, which had begun in earnest during March, slowed as temperatures shifted to neutral. The most rapidly warming area in the Pacific at the time was in Niño 3.4. It should be noted that the warmth did not center in the latitude of the equator at the date-line then, so it really was not an El Niño event in the classical sense. It was an unusual warming of the extra-tropical Pacific. The Sun had moved across the equator in March, and the ocean to the north of the equator was filling with warmth but the Kelvin wave activity that characterizes an El Niño event had not taken place. Kelvin waves are large eastward-propagating ocean swells that move from the dateline toward Peru starting in June or July of an El Niño year. The large, slow-moving waves carry warm water along the equator and serve as the driving force of an El Niño. The generation of these waves is more of a summer pattern in which storms at the dateline can start waves that drive the warm surface water toward the coast of Peru. The warming at this time was not a full Kelvin wave pattern. It was just a prelude.

In the middle of May 1982 (next image), Mars went direct (eastward) in western Niño 4. This motion was coincident with a brief, slight fall of SSTs to the west, even though the temperatures were actually above average. Typically, at this time, when there is much warmth in the western Pacific, a cooling trend sets into Niño 4 as the warm water begins to drift to the east. In May 1982, this canonical feature resulted in a neutralizing western sector of Niño 4, even though the SSTs in that area remained above average. However, this shift was coincident with the establishment of a strong warmth center at the dateline and the southward migration of the mass of warm water that was extratropical into a tropical latitude. The equator was now the site of strong anomalous warming patterns. Mars and the canonical El Niño appeared to

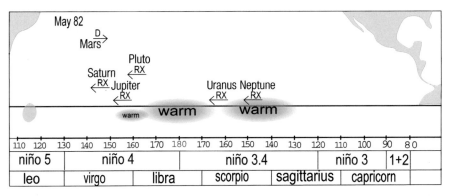

be synchronous with this shift. However, the turning of the tide was really just getting under way.

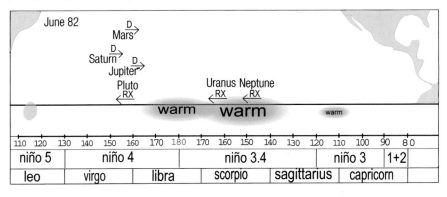

Jupiter and Saturn also went direct in Niño 4 in mid-June 1982. This motion was coincident with a rise in SSTs in Niño 4 and at the dateline. Warmth in the far western Pacific had neutralized, and any cold in the east was long gone. The action was in the space between the cluster of planets in Niño 4 and the two planets in Niño 3. By the end of June, Mars, Jupiter, and Saturn were moving eastward in the western Pacific, and the dateline was showing signs of a strong warming, as well as the regions near Neptune in Niño 3.4. Neptune and Uranus were still in retrograde motion opposing the eastward movement of Saturn, Jupiter, and Mars. This created a squeeze pattern in Niño 3.4 coincident with the strong rise in temperatures in that area.

Very early in July, Pluto went eastward direct in Niño 4. This meant that, at the very beginning of the important El Niño month of July, four planets were moving eastward toward the all-important dateline area in the mid-Pacific.

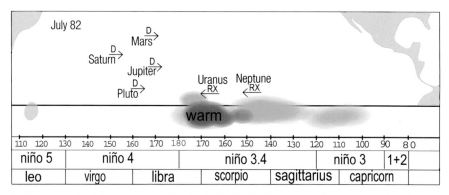

This was coincident with a rise in SSTs in Niño 3.4 to the east of the cluster. The reason these configurations are important is that the conditions at the dateline in July determine the outcome of the El Niño for that year. In this model, it has been found that a situation in which planets are moving toward the dateline in June and July is often coincident with enhanced El Niño events in December.

An added feature in the summer of 1982 was that the convoy of eastward-moving planets in the west would form a squeeze pattern with the westward retrograde movements of Uranus and Neptune. This squeeze was coincident with a strong warming of eastern Niño 4 and Niño 3.4. This warming pattern was coincident with the development of the warmth flow in the Pacific focused around the equator and between the dateline and the western edge of Niño 3 in July 1982.

By early August 1982, Jupiter had moved into eastern Niño 4 and was crossed by Mars as it moved rapidly into western Niño 3.4. This was a critical Mars movement at the dateline at a critical time in the canonical Niño year. This motion was in the context of the eastward movements of Jupiter, Pluto, and Saturn in direct motion in Niño 4. These combined eastward movements were coincident with a tremendous warmth flow from the dateline toward Hawaii at the time. In early August, at the exact time when Mars was crossing Jupiter just to the west of the dateline, Uranus in western Niño 3.4, just to the east of the dateline, went direct (eastward). This happened at the precise time that Mars was passing Jupiter and approaching Uranus from the west. This type of pattern in this model is known as a "slingshot" because of the tendency for energies to be radically compounded when this type of pattern is involved. The sudden intensification of warmth in the eastern tropical Pacific was remarkable, both for its intensity and for the coincidental planetary timing.

To the west, Mercury also entered the picture by coming into western Niño 4 late in the month. As these movements unfolded, a strong cooling impulse emerged in the far-western Pacific as the warmth moved to the dateline area and beyond. In Niño 3.4, the center of the strongest warming was the squeeze area between Uranus and Neptune. This is in line with the model.

110	120	130	140	150	160	170 180	170 160 150	140 130	120 110 100	90 80
niño 5		niño 4					niño 3.4		niño 3	1+2
leo		virgo		libra		scorpio		sagittarius	capricorn	

At the very beginning of September, Neptune went direct (eastward) to the east of Hawaii. This motion was coincident with the development of a moderate warmth tongue arising on the west coast of Peru. Now all of the planets over the Pacific were moving in direct motion. The center of the warmth had shifted with the Neptune motion and was now in Niño 1+2 and Niño 3. The western Pacific was cool as Mars moved across the dateline and into the eastern Pacific. At the very end of September, Venus entered Niño 4, and, at the same time, Mercury went into a retrograde loop in a squeeze pattern with

Venus. This movement was coincident with the dissolving of the growing cold pool in the western Pacific, bringing it into a more neutral condition. The warmth in the east diminished slightly as Mercury went retrograde in the west in the latter part of the month.

At the beginning of October, Mars passed Uranus in Niño 3.4. This was coincident with an enhancement of the warmth tongue in Niño 1+2. In mid-October, Mercury once again went direct in Niño 4. All of the planets over the Pacific were now moving in direct motion. The pool of warmth against the west coast of Peru was again becoming very active in sending out a warmth tongue along the equator toward the dateline. By the end of the month Mars had passed Neptune in Niño 3.4. This was coincident with an extreme burst of warmth from the Peruvian coast that spread along the equator moving west at a rapid pace. In the west, Mercury and Venus were now running in tandem with each other near the dateline.

In November 1982 (opposite top), no planets changed direction other than the constant eastward motions of Mars in the east and Venus and Mercury streaming past the dateline. The extreme warming pattern set in motion in October in the eastern Pacific continued unabated throughout November. By the end of November, Venus and Mercury were neck and neck in Niño 3.4, while Mars had reached Niño 3. SSTs in the eastern Pacific, at the border between Niño 3.4 and Niño 3, were raging as the warmth tongue from Peru extended westward to the dateline.

By mid-December, Mercury, Venus, and Mars were all in Niño 3. By December 25, Mars had reached the western border of Niño 1+2. By the end of December, the trio had reached Niño 1+2. The timing of these movements

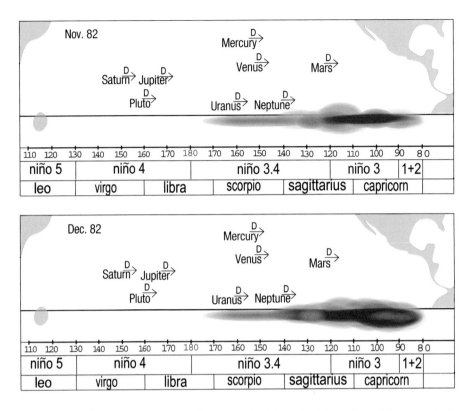

was unusually precise. It was also remarkably coincidental to this record El Niño event. Starting in 1981, the series of loops and squeeze patterns shows remarkably coincidental relationships between planetary movements projected onto the Pacific basin and the climatic phenomenon of the El Niño.

TRACKING THE RECORD EL NIÑO 1997/1998

In the modeling procedures used to research El Niño and La Niña events, the motions of the inner planets over the Pacific are often a useful tool for seeing into the formation and migration of warmth pools from the dateline. Two patterns are useful for the observer. The first is a block, which occurs when a fast-moving planet—Mars, Venus, or Mercury—goes into apparent retrograde motion between the international dateline and the West Coast. In a block, there is often a brief rise in temperatures in the longitude of the retrograde motion. If no other planets are involved, temperatures usually go back down after the retrograde motion has ceased.

squeeze patterns

1991	1996
Mercury	Mercury
D RX	D RX
Mars	Venus
warm	warm

110	120	130	140	150	160	170	180	170	160	150	140	130	120	110	100	90	80

niño 5	niño 4	niño 3.4	niño 3	1+2
leo	virgo	libra	scorpio	sagittarius	capricorn

When another planet is approaching from the west or is situated in the west and moving in direct motion, a planet forming a loop to the east confronts the approaching planet from the west. This pattern is known as a "squeeze." In squeeze patterns, it may be that the planet approaching from the west is accompanied by increasing temperatures in the zone it is transiting. Usually, the squeeze is coincident with a sudden leveling-off of the rising temperatures, but the temperatures remain at elevated levels after the squeeze is completed. A squeeze between Mars and Mercury in Niño 3.4, in November 1991, was coincident with a turning point for the anomalous warming that took place in the spring of 1992 in Niño 1+2. In 1996 a squeeze between Venus and Mercury in Niño 3 took place at a time when SSTs in Niño 3 were below average and headed downward. The time of the squeeze was coincident with an abrupt turnaround in temperature trend for the eastern Pacific. This turnaround was the initial event in a series of events that would result in the great El Niño of 1998.

In early December 1996, a cool eastern Pacific began to change temperature trend toward a warming phase. Jupiter, Uranus, and Neptune in Niño 3 were in direct motion. Jupiter was rapidly approaching Neptune for a conjunction in Niño 3. Around Christmas 1996, Mercury went conjunct to Jupiter and simultaneously went retrograde in Niño 3. Venus also supported this strong squeeze in direct motion in Niño 3.4. Here was a complex array of motions in one sector of the Pacific. The close proximity of the Mercury retrograde motion to the Jupiter–Neptune conjunction brought together many opposing forces into a small time–space configuration. These complex movements were coincident with a truly unusual warming impulse in the eastern

Dec. 96 — Jupiter D, Uranus D, Neptune D, Venus D, Mercury RX, warming

110 120	130 140 150 160	170 180 170 160 150 140	130 120 110 100	90	80
niño 5	niño 4	niño 3.4	niño 3	1+2	
leo	virgo	libra	scorpio	sagittarius	capricorn	

Pacific. This warming changed the trend in the eastern Pacific. At the same time in the west, Mars was entering the western Pacific, where neutral conditions were established.

Jan.-Mar. 97 — Jupiter D, Mars RX, Pluto RX, Uranus D, Neptune D, warming warming warming

110 120	130 140 150 160	170 180 170 160 150 140	130 120 110 100	90	80
niño 5	niño 4	niño 3.4	niño 3	1+2	
leo	virgo	libra	scorpio	sagittarius	capricorn	

The year 1997 began with a pool of warm water in the western Pacific and a neutral but warming trend in the eastern Pacific. In February, Mars went retrograde at the western border of Niño 4. This motion was coincident with an enhanced warmth buildup in the western Pacific. This motion was a perfect example of support for the buildup phase of the canonical El Niño. In March, Pluto went into retrograde motion in Niño 3.4 near Hawaii, further supporting the canonical El Niño as warmth continued to move SSTs toward neutrality in the western Pacific. SSTs from the dateline east were below average but slowly building toward neutrality in the area between the planets that were moving in opposite directions.

In April, in Niño 4, Mars shifted into direct motion. Pluto was retrograde in Niño 3.4. The space between them was a squeezed area at the

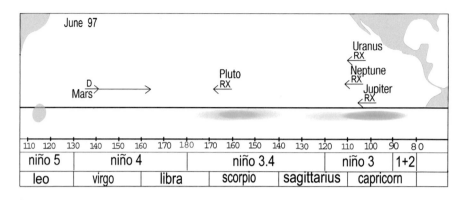

dateline. The dateline in April was the site of a strong warming impulse. In May, Neptune and Uranus were moving in retrograde motion on the border between Niño 1+2 and Niño 3. This motion was coincident with a strong warming impulse spreading from Niño 1+2 into Niño 3. A warmth plume started to emerge from the west coast of South America at the equator. Warm water was moving to the west of the planetary cluster. The situation in early May was that there were two pools of anomalous warm water in the equatorial Pacific, one at the dateline and one in the eastern Pacific. The western pool was following the canonical El Niño with a slow drift toward the dateline. The eastern pool, under the influence of Jupiter, Uranus, and Neptune, was in direct motion toward the west coast of South America. This motion was coincident with a strong warmth plume coming off the coast at the equator and streaming toward the dateline. On May 1, Neptune went into retrograde motion, and the anomalously warm tongue at the equator sputtered. On May 13, Uranus went into retrograde motion, and the anomalous warmth plume died off. What happened? According to the protocols used in this model, when a planet moves retrograde the warmth appears to the west, but cold appears to the east. In this scenario, the retrograde motion of the two planets cooled the anomalous warmth plume that was to the east of their positions. This was further enhanced when, on June 10, Jupiter went into retrograde motion in Niño 1+2. At the beginning of June, the impressive anomalous warmth plume from South America, which is the signature of a robust El Niño, was not to be seen. Of course, the eastern Pacific was warming at this time, but the anomalous warming that took off in May diminished in the exact time frame of the shift in Neptune and Uranus in May 1997. This anomaly was suppressed further with the retrograde motion of Jupiter in early June. By late June, however, with Uranus, Neptune, and Jupiter all moving westward in the eastern Pacific, the anomalous warm tongue once again began to appear at the equator. It seems the cold influx had run its route, and now it was once again time for the anomalous warmth to appear.

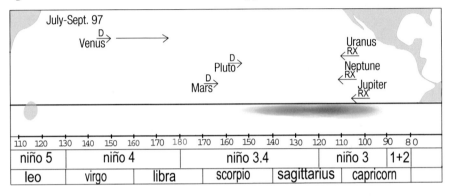

From July to mid-September, there were few planetary changes in direction over the Pacific—except, that is, for the constant eastward motion of Mars, which brought it to the longitude of Hawaii and Pluto by September. Pluto had gone direct near Hawaii in August. The trio of Jupiter, Uranus, and Neptune in the eastern Pacific were still moving in retrograde motion, and the anomalous warmth plume was still intact along the equator, from Peru to the international dateline, in the space between all of these planets. In August, Venus moved into Niño 4 to begin its Pacific transit for the year. In September, Venus crossed the dateline.

In early October, Mars crossed Pluto with both planets moving direct near Hawaii. Since there already was a large pool of warmth in this area, the crossing was coincident with a migration of warmth into Niño 3.4. This was the beginning of a squeeze pattern, which was centered on the area between Pluto, and Mars in the west, with Jupiter, Neptune, and Uranus in the east. In late September, this area was the site of unusual warmth in the ocean at the equator. Then, between the 8th and the 14th of October, the cluster of planets in the east went direct. This meant that in 1997, from October 14 onward, in the eastern Pacific all planets were direct with a large area of warm water between Hawaii and Peru. Anomalous warmth accelerated at the equator in October and November. In November, Mars continued its transit to the east and reached the cluster of Jupiter, Neptune, and Uranus in the critical month of December. In the west during October, Mercury made an appearance in Niño 4 to begin its Pacific transit for the year. In a truly screaming transit, Mercury moved from the western side of Niño 4 to the eastern portion of Niño 3.4 in two months. Venus, at a more pedestrian velocity, had moved from the mid-Pacific to the eastern Pacific, crossing Hawaii in October.

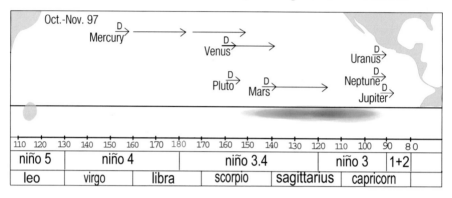

Since September 1997, the media made dire predictions regularly about the scope of this El Niño. By late November, however, the much-awaited event was not manifesting according to the models. Anomalous Sea Surface Temperatures were present at the equator, but none of the other phenomena associated with an El Niño year seemed to be unfolding. By December, scientific sources actually published bulletins that expressed doubt about the emergence of an El Niño that winter. One expectation was for unusual rains on the West Coast of the U.S., but through November the storm jet dropped south on the coast only weakly and intermittently. By December, there was an enormous pool of warm water off the coast of California, and yet no storms arrived on the coast. This hesitation of a record El Niño was coincident with the simultaneous retrograde loops of Mercury and Venus in Niño 3 and Niño 3.4. These simultaneous blocks created a strong retro pattern on the West Coast at the time. Mercury was to the west, so its influence was not as great. However, Venus went retrograde in the middle of the cluster of all of the other planets gathered there against the coast. This was a squeeze of epic proportions.

The El Niño did not break on the coast until after Mercury went direct in late December. Skies clouded and curious, extremely light rains developed,

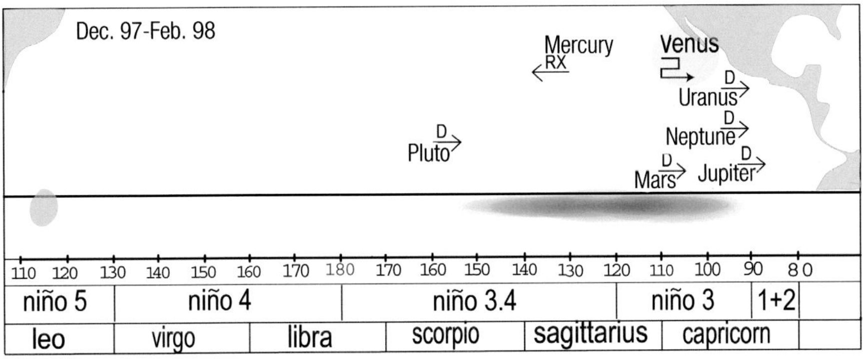

Dec. 97-Feb. 98

Mercury RX ← Venus

Uranus D →

Pluto D →

Neptune D →

Mars D → Jupiter D →

110	120	130	140	150	160	170	180	170	160	150	140	130	120	110	100	90	8 0

niño 5	niño 4	niño 3.4	niño 3	1+2
leo	virgo	libra	scorpio	sagittarius	capricorn	

but no true El Niño phenomena. Then, when Venus finally went direct in February, the full force of the El Niño pattern emerged. At that exact time, the storm jet dropped south on the West Coast, and record rains were the result.

This points to the fact that, even when the physical conditions are perfect for an El Niño, there is often a planetary perspective that can shed some light on the most unusual and perplexing climatic events. The El Niño of the century did not manifest while Mercury and Venus were retrograde in the eastern Pacific. That El Niño event broke into the record books when the blocking and squeezing retrograde planets finally went direct.

Rain on the ocean (by Fedorov Oleksiy, shutterstock.com)

CHAPTER 4

EARTH LINKS

The task of meteorology is to see daily weather patterns a few days before they arise. The task of climatology is to see seasonal weather patterns over larger periods. Most people consider it impossible to predict weather patterns a year or two in advance. In fact, however, it is possible to see more deeply into periods using a modeling protocol that involves planetary-motion data. To accomplish this, charts were eventually developed that allow the tracking of weather patterns at the synoptic or hemispheric level. This book provides a various approaches to forming a synoptic long-range climate model. The illustrations on the following pages depict the most fundamental features of the research charts used in this system.

In the yearly round of events, research has shown that the most singular event for the Earth Soul is the phenomenon of the eclipse. For many years, numerous case studies have made this evident. Using eclipses as a beginning to building planetary linkages has gradually resulted in the construction of an eclipse grid that allows the projection of planetary motion data onto the Earth in specific longitudes and latitudes. The first figure (page 118) illustrates the positions of the eclipse points and their reflex points. The sets of points shift to the west with each new eclipse. Each new position provides a series of lines to arise and form an eclipse grid. The eclipse grid has proved effective in predicting the placement of highs and lows for the six-month period between eclipses.

The chart shows the two main eclipse lines going from their respective eclipse points to the reflex points. One line goes 180° from the solar eclipse point to the solar reflex point, and another line goes 180° from the lunar eclipse point to the lunar reflex point. In the eclipse grid model, there is little difference considered between an eclipse point and an eclipse reflex point. They both have shown to have similar values and actions. The reflex points are the positions exactly opposite the solar position on the day of either a

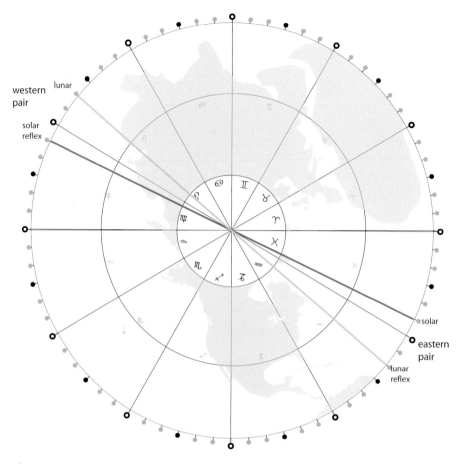

lunar or a solar eclipse. Although no clear reason for the effectiveness of reflex points has emerged from the research, reflex points have proved to be as powerful as the actual positions of the eclipses themselves.

The solar and lunar points to the east are known as the "eastern pair," while the solar and lunar points to the west are the "western pair."

In the next figure (opposite), a set of jet curves have been added to the eastern pair of eclipse points. Jet curves are harmonic curves that describe particularly sensitive areas in relation to the position of an eclipse point. Through established protocols, the jet curves can be projected onto areas expected to be climatically significant according to the current eclipse placement. This chart indicates the position of the lunar node between the two eastern eclipse points. The lunar node has the quality of a planet, in that it can seriously

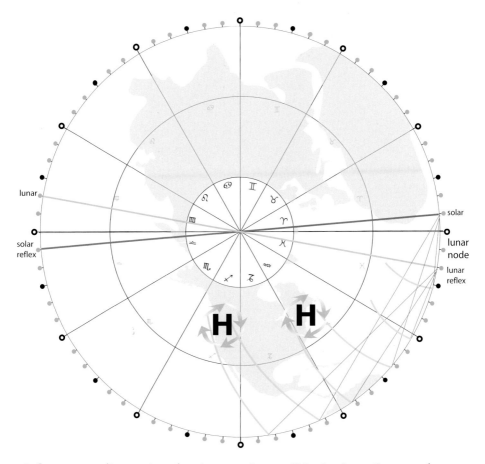

influence an eclipse point when it moves in arc. "Moving in arc" means that a
planet, or in this case the lunar node, had moved celestially into a new degree
of longitude. This information comes from an ephemeris that provides the
motion-in-arc data for each planet every day. Motion-in-arc data is the heart
of this system. In practice, the rhythmic cadences of the motion-in-arc increments are projected to each eclipse point for each motion in a day so that the
charts used for the forecasts can be constructed. Different geometric combinations create the potential for high or low pressure on the jet curves.

In this chart, the motion of the node has placed a high-pressure value on
to both of the eclipse points. To the west of the points, the jet curves over
the continent are showing the effects of that high-pressure influence in the
formation of a high over the Maritime Provinces and another over the Pacific

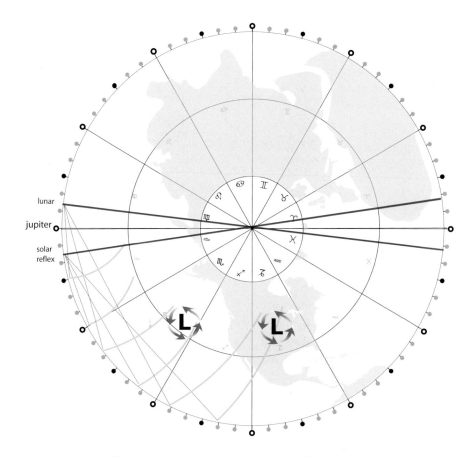

Northwest. The projection of jet curves from eclipse points across the northern hemisphere is the fundamental technique in this modeling approach. A planet or lunar node moving in arc near to an eclipse point consistently influences the activity of troughs or blocks within harmonically distributed zones, either to the east or to the west of the eclipse point. These disturbance zones often center on the jet curves.

The placement of the jet curves in specific years has proved effective in determining the fundamental climatic scenarios for the particular season. Sometimes the jet curves are over significant climatic areas, and sometimes they are not. The various positions can model the potential for many different climatic scenarios. The potential for the jet curves to support high (ridge) or low (trough) formations has a robust predictable influence on long-term weather events. In the chart on page 119, the placement of the two ridges would cause us to expect that the area near Nevada and the Northeast would be centers of high pressure, with a potential for troughs to form in the upper Midwest between the highs. Reliable long-range forecasts are made from these modeling procedures.

In the third figure (opposite), Jupiter is influencing the jet curves from the western pair of eclipse points. This influence is for low pressure rather than the high-pressure influence we saw from the node. These influences shift and change rapidly in time and constitute a constantly shifting web of potentials in the atmosphere. In the chart, we see that Jupiter's influence on the eclipse points is creating a tendency for low pressure over the eastern Pacific and over Nevada. From this placement, we might expect storms into the West Coast and Pacific Northwest as well.

In the fourth image (page 122), we see the crossing point of two of the jet curves within in the zone labeled as the disturbance diamond (yellow). The crossing jet curves are placed at 72° angles of arc from the eclipse point that generates them. They are 72° jet curves. However, they cross in the area dominated by a zone that includes a combination of the jet curves from both pairs of eclipse points. The crossing of the curves creates the disturbance diamond (yellow). This zone tends to be the most turbulent and decisive area in any given chart, since a planet influencing the western pair to one value and a planet influencing the eastern pair to the opposite value will generate these opposing values on their respective 72° jet curves. The two 72° jet curves with the opposing values will cross their influences in the disturbance diamond. This zone gets its name from the zone that arises where the two 72° jet curves cross.

The chart depicts both the high- and low-pressure areas coming from the opposite eclipse pairs. Of course, we could reverse these values, with the node creating low-pressure values on the eastern pair, and Jupiter creating high-pressure values on the western pair. The interesting aspect of this patterning is that the shift of values between the two planets might happen on the same day, sending an established pattern that had been prevailing for a few weeks into a state of chaos. This type of event is the truly fascinating element of the eclipse-grid model. When working with the eclipse grid, one or two planets normally dominate the space around one or the other set of eclipse points. The disturbance diamond zone then becomes the dominant area to watch when planets

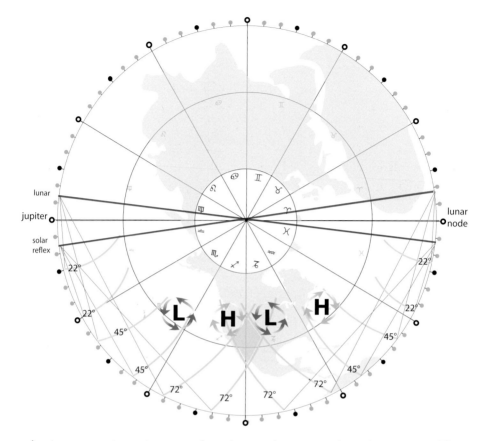

begin to move in arc in unusual or abnormal sequences in a given season. The eastern and western pairs of eclipse points respond to the motions of planets approaching and moving away from them.

In this image, other jet curves from each point are also included. Besides registering on the 72° jet curves, planetary motion responses are also registered on 45° jet curves and 22° jet curves. These curves are resonant areas that are remote from the eclipse points but link to them harmonically.

To make a forecast, the planetary motions around the complete set of jet curve zones and their accompanying values of high and low pressures are woven into the fundamental climatic scenario for that season in that locale. In the eclipse-grid model, these patterns are then compared to analog years when the eclipse points were in similar positions, and a forecast is made based on the climatology or unusual weather patterns of the analog year.

Planet approach and retreat patterns

In the language of astronomy, a transit occurs when one planet crosses the face of another planet. In the language of astrology, a transit occurs when one planet moves in arc in the general longitude of another planet. In the eclipse-grid model, a transit occurs when one planet strikes a particular angle to an eclipse point that generates a high- or low-pressure value on the jet curves linked to that eclipse point. When tracking the transiting motion-in-arc of a planet in relationship to a fixed eclipse position, a wave-like rhythmical pattern of high- and low-pressure values emerges. A moderately moving planet such as Jupiter, when moving near an eclipse point without interference from any other planet, will reveal a Doppler-like wave effect of high- and low-pressure changes on the jet curves. The shifting numerical values generate responses on the jet curves in the eclipse grid. The forces on the jet curves then interact with the dominant climatic regime over which they pass. The Doppler wave unfolds first as a planet approaches the point, then as the planet conjuncts the point, and finally as the planet moves away from the eclipse point. This figure shows a wave of high- and low-pressure fluctuations as a wave of oscillating high and low pressures. The numbers represent the angular aspect in degrees of arc through which a planet moves as it approaches a conjunction with an eclipse point. A 10° angle of arc refers to the position of a planet when it is 10° from conjunction with an eclipse point.

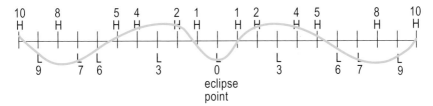

We see in this figure that a planet approaching an eclipse point from 10° to 6° would stimulate predominantly low-pressure values. The exception would be the high-pressure position at 8°. The same planet continuing it's approach would generate high-pressure values as it passes from 5° to conjunction at 0°, the exception being when the planet passed 3°. At the conjunction itself, a strong low would be generated in the period of the conjunction. On the other side of the conjunction, the mirror pattern would unfold.

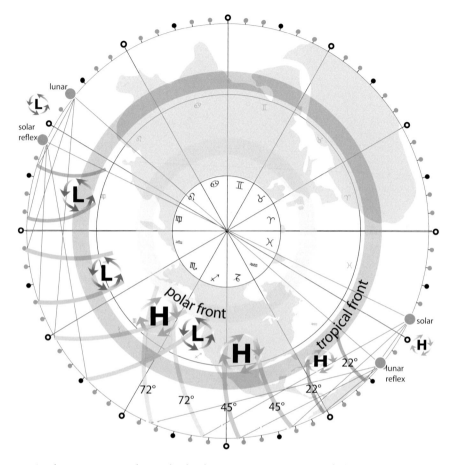

A planet moving through the low-pressure section of the Doppler curve would engender a low-pressure response on the 22°, the 45°, and 72° jet curves projected from that eclipse point (above). Low-pressure values create troughs on the jet curves. By contrast, high-pressure values on the eclipse point tend to create ridges on the jet curves. The chart shows a situation in which the Atlantic or eastern pair of eclipse points is influenced to high pressure. All of the jet curves from that point have a ridge over them. There is a ridge over Bermuda on the 22° jet curves, a ridge over Denver on the 45° jet curves, and a ridge over the Gulf of Alaska on the 72° jet curves.

As a planet passes through the series of angular aspects that stimulate low-pressure values on the eclipse points, low pressure shows up on the jet curves. The western pair of eclipse points has low-pressure values associated with it.

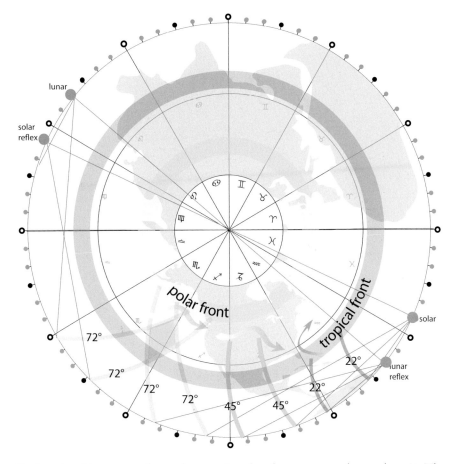

Each pair of jet curves from those points has low pressure located on it. The West Coast of the U.S. is experiencing the passage of a trough as a result of this configuration, as is the central Pacific on the 45° jet curves. The same jet curves would shift to high pressure when the planet continues into the high-pressure side of the Doppler curve moving from 5° to 1°. The jet curves associated with the points aspected by a transiting planet would then manifest ridge formation, as well as trough formation in time frames coincident with the approach, conjunction, and retreat of the transiting planet.

In the figure above, we see a double set of 72° jet curves projecting from each eclipse point. These curves cross each other to form the previously mentioned disturbance diamond (yellow). Being the center of so much geometric focus, this nexus of 72° lines often makes it the dominant weather maker in a given chart, especially when the area is placed over a sensitive climate zone. In the position illustrated, the disturbance diamond would be problematic in steering storms from the polar front into the Pacific Northwest. The polar front is the prime source of winter storms. It is often entrained by aspects creating ridges on the disturbance diamond.

However, in a given year it may be that the 45° jet curves are over a particularly sensitive climate area. Then, planetary motion events near to the eclipse points would most likely manifest on the 45° jet curves (green). Those curves tend to influence mid-latitude phenomena. Air masses linked to these curves rarely surge into the high-latitude polar regions, but they are often influential in sensitive climatic niches like the Midwest, where frontal passages are often the result of mid-latitude influences (green arrow). The low-latitude 22° jet curves (pink) have been the most difficult to document, since their influences often emerge at either a very high latitudes (Greenland blocking patterns in the polar front) or very low latitudes such as hurricanes moving north out of the tropical front (pink arrow). Case studies show that the effectiveness of the 22° jet curve influences are linked strongly to specific, climatologically sensitive areas. The study of 22° jet curve influences is an ongoing process.

The final chart (opposite) shows the double set of eclipse points projecting the mid-latitude 45° jet curves in three different positions. This shows the typical migration of the set of eclipse points through time. The red points project a set of 45° jet curves onto the U.S. East Coast. If a planet such as Jupiter is in a high-pressure relationship to those points, a ridge will form over the East Coast. That would indicate warm weather for the Great Lakes and Northeast. The next year or so would find the eclipse points in the green position. A ridge projected to the pair of 45° jet curves would form over the inter-mountain area, bringing warmth to the desert Southwest, but strong cold to the High Plains and Central States. A few years later, the blue eclipse pair would project mid-latitude 45° jet curves into the eastern Gulf of Alaska. A ridge there would bring turbulent cold weather to the Great Basin and PNW in what is known as an *inside-slider pattern* where fronts from Alaska move through BC on their way to Utah. This shows how the eclipse grid moves through time and how the influences shift from year to year

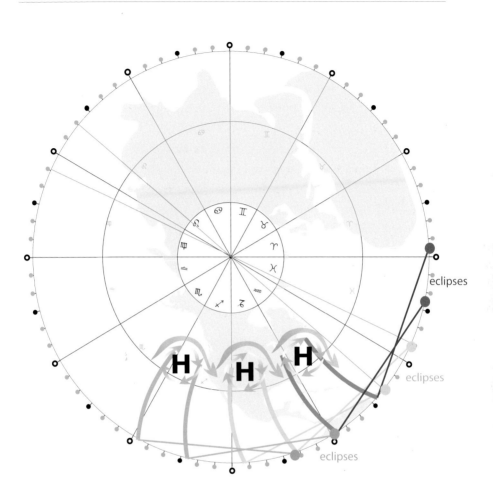

In years when the area between the 45° jet curves is the site of strong planetary effects, they can be the dominant mid-latitude element in the chart, especially when the area covered by the two jet curves is situated in a climatically sensitive area such as the Hudson Bay or Greenland or over Vancouver Island, British Columbia. In such years, mid-latitude areas dominate the climate patterns. The 45° jet curves have a strong effect on the mid-latitudes and can be used most effectively to reckon monsoon influences and the extent of the mixing potentials between the polar front and the tropical front, as planetary motions approaching and retreating from the eclipse points exert their influence on mid-latitude storm tracks in a given season.

THE LUNAR NODE

In astronomy, *declination* (Gr. = decline, below) is the angular distance of a celestial object measured north or south of the celestial equator. *Angular distance* is reckoned in angles of arc on a sphere. Imagine a point on the equator. That point on the equator would have a 0°angle of arc from itself. A point that would be found at 90° angle of arc in north latitude from that point on the equator would be at the North Pole. The point at the North Pole would actually have 0° angle of arc to its own position. As a result, arc angle is the distance on a sphere from one point to another. Arc angles north and south are measurements of latitude and arc angles east and west are measurements of longitude.

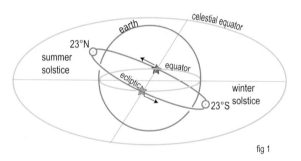

fig 1

In figure 1, the blue circle represents Earth, and the smaller green ellipse the equator. The solar orbit is in red, and the celestial equator is seen as a projection of the terrestrial equator. The apparent movement of the Sun seen from Earth moves in a counterclockwise path called the "solar orbit." We see that the Sun does not follow the celestial equator in its apparent orbital path. The red star represents the point where the sun will cross the celestial equator moving from south to north in spring. The red star is the spring, or vernal, point that designates the position of the Sun on March 21 at the spring equinox. The spring point is important in the seasonal year. In many cultures, it symbolizes the renewal of spring and has often been considered full of import about the qualities of the following year. From the vernal point of the spring equinox, each day the Sun rises higher in the sky at noon until, at noon of the summer solstice, it appears at 23°N latitude. Continuing from the highest point in declination at the summer solstice, the Sun then appears to drop in celestial latitude each day toward the equator until September 21 and the fall equinox. The Sun is again on the equator (green star), gradually moving south each day in celestial latitude until, at the winter solstice, December 21, the noontime Sun appears at 23°S celestial latitude. When the Sun is below the celestial equatorial plane in a southern declination, it is winter in the northern hemisphere. The passage of the Sun in declination above and below the plane of the celestial equator is the source of the seasons.

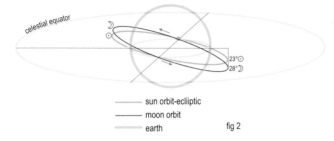

——— sun orbit-ecliiptic
——— moon orbit
——— earth   fig 2

The rotation of all of the planets on their axes is from west to east, or counterclockwise (arrows in figure 2). As a result, the Sun rises in the east and sets in the west, as observers on the Earth rotate from the west toward the rising Sun in the east, passing under it at noon, and then seeing it set in the west as they continued on in their journey counterclockwise to the east. The orbital motion of planets around the Sun is also from west to east, or counter-clockwise. Seen from Earth, the Moon, moving day after day through its orbit, slowly passes counterclockwise, from west to east, by increments of 13° of arc each day (black ellipse). This motion is experienced from the perspective of the Earth as relative to the fixed stars behind it. An observer on Earth would see the Moon moving as if it were on the edge of a great circular plane. This is the orbital plane of the Moon around the Earth.

In figure 1, we see that, if an observer extended the equator of the Earth toward the heavens, it, too, would become a plane—the plane of the celestial equator. The stars and planets below this plane would be in the southern celestial hemisphere, and the stars and planets above this plane would be in the northern celestial hemisphere. In the first image, we see how in the course of a year the Sun appears to go above and below the plane of the celestial equator (red ellipse).

One of the most eccentric of orbital planes is that of the Moon (figure 2, black ellipse). The lunar orbit is 5° eccentric to the orbit of the Sun. That makes the maximum southern and northern declination of the Moon 28°.

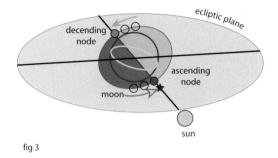

fig 3

Each month ("Moon-th"), the Moon mirrors the entire yearly orbital journey of the Sun. Each month, the Moon passes through the southern portion of its orbit for two weeks. The darkened portion of the central ellipse in figure 3 depicts this. The Earth is the blue circle with the equator as the green ellipse inside it. The purple–orange ellipse shows the orbit of the Moon around the Earth. The large, tan ellipse is the orbital plane of the ecliptic of the Sun. The diagram shows the motion of the Moon as it passes through the ecliptic plane of the Sun each month as it orbits the Earth. In this orbit, the Moon moves counterclockwise from south to the north for two weeks. The small red circle shows the motion of the Moon toward the place where the last eclipse occurred (purple star). This point, where the Moon crossed the ecliptic for the last eclipse, is known as the "ascending node," where the Moon passes from the southern celestial hemisphere northward across the apparent plane of the solar orbit or ecliptic plane (tan ellipse). The moon is moving eastward (counterclockwise) at a rate of 13° of longitude each day (red arrow). At the ascending node, the Moon moves across the solar path and into the northern celestial hemisphere. The orange section of the central ellipse shows this two-week passage. For two weeks, it is high in the night sky mirroring the path of the Sun in the northern hemisphere summer. Two weeks later, it crosses the ecliptic again at the descending node (green Moon). This time the Moon is moving into the southern celestial hemisphere. Every six months, through the immense mystery of orbital rhythms, the Moon crosses the ecliptic just as the Sun is at the same spot on the ecliptic. When this happens, the Moon can

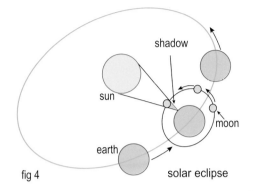

fig 4   solar eclipse

cast a shadow on the Earth and a solar eclipse occurs. Two weeks later the Moon is on the opposite side of the Earth from the Sun, and the Earth casts a shadow on the Moon during a lunar eclipse. For either eclipse to happen, the Earth and Sun and Moon must first line up with each other on the ecliptic. This creates an eclipse as the two planets, seen from the Earth, are directly in line with each other.

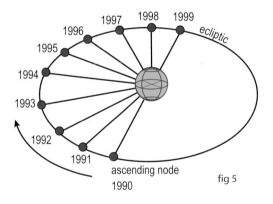

fig 5

The nodal points where the Moon and Sun intersect every six months are not stationary. Because it is elliptical, the lunar orbit has two centers, and the oscillation of these two centers around the North Pole creates what is known as a precession of the nodal positions. *Precession* means that each eclipse is a little progressed, or precessed, to the west of the preceding eclipse.

The precession of the nodes has a period of approximately 18.6 years (figure 5). This is a nodal cycle. The east-to-west motion of the nodes means that a regular progression of the eclipse points is constantly migrating backward against the counterclockwise motion of the other planetary orbits. The important point for climate phenomena is that, as the nodes precess, the positions of the eclipses move through the zodiac in periods of 9.3 to 18.6 years. These periods are very close to the well-known climatology principles of the decadal (10 years) and inter-decadal (20 years) return cycles of climate patterns.

The next image (opposite) shows the motion of the two eclipse points between 1999 and 2005. The solar eclipse in January 1999 is at the position of the solar eclipse at 5 p.m. on a clock face. This position is read as 2° Aquarius (AQ). Across the circle from that position at 15° Cancer (CA) is the placement of the lunar eclipse also in January 1999. The pink segments show the distance in arc between the solar and lunar points in January 1999. The next set of eclipses occurred six months later in July 1999. Those are depicted by the lunar eclipse at 9° Capricorn (CP) and the solar eclipse in July at 23° CA. The space between these two eclipses is represented by the dark blue bars. It can be seen from the chart that the movements of the nodes where the eclipses occur are progressing from east to west. The earlier eclipse covers an area over the Mississippi Valley in North America, and the later eclipse covers an area over the Plains states of North America. This is the motion of the nodes.

We see from the chart that the nodal motion is far from consistent. There are gaps and congestions in the sequences of the progression of the nodes across the land masses of the Earth. Sometimes the nodes seem to be stuck in positions for a year or so, and at other times they seem to spring far ahead of the previous positions. Sometimes the lunar eclipse is first in the sequence; other times, the solar eclipse is first. There is a rhythm of about four or five years between the times of these reversals. This chart is included to show how the motion of the nodes is a constant rhythmic pulse that keeps the accompanying eclipse grid moving through the seasonal changes and climate events. The phenomena of the climate have an intimate relationship to the rhythms of these lunar and solar periods.

eclipses 1999-2005

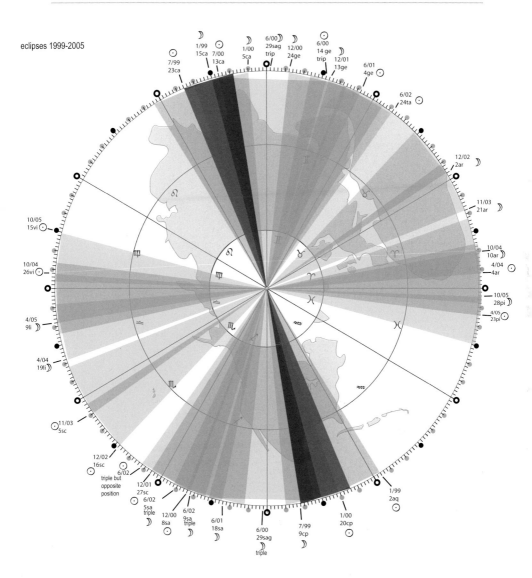

MERCURY'S INFLUENCE ON WEST COAST JET-STREAM PATTERNS

The orbital motion of a planet moving against the background of the fixed stars is known as "motion-in-arc," a very useful tool for studying climate anomalies. The key is to find a phenomenological protocol for showing the coincidence of motion-in-arc events of a planet to particular climate phenomena. To be useful, the linkage not only has to depict coincidence of time frames, but also it must show that particular planetary motion-in-arc events were linked to particular climate phenomena in particular geographical locations.

The following study is given as a suggestion for such a phenomenology.

Through repeated observations, it has been possible to identify regularly observable planetary movement patterns connected to fluctuations in the polar jet stream in the northern hemisphere as the passage of a planet through a particular longitude. Earlier chapters gave examples of the responses in air masses to the passage of the Moon across the West Coast. It has been repeatedly observed that, as the Moon passes by the longitude of North America, air masses that have been stationed in a particular longitude will respond to the passage of the Moon by moving in the direction of the lunar passage in the exact time frame of the transit. The motion in longitude of a planet transiting a given longitude is often coincident with the motion in longitude of a stationary air mass in that longitude. It is a kind of tidal effect in the air that is readily observable.

Related to this passage in longitude is an accompanying movement in latitude of a transiting planet. The rapid motions of the Moon make its longitudinal passage more perceptible. It is more difficult to study phenomenologically other planets that are moving more slowly than the Moon does. A reference is needed to see the effects of slower planets. The Sun moving in longitude and latitude is a good reference.

The following image shows the path of the Sun's ecliptic projected onto the surface of the Earth. In the chart, the green waveform represents the apparent path of the Sun as it moves from the northern hemisphere into the southern hemisphere and back again over the course of a year. The Sun moves one degree of longitude in one day as it travels from west to east across the equatorial regions. In the image, we can see that the path also goes above

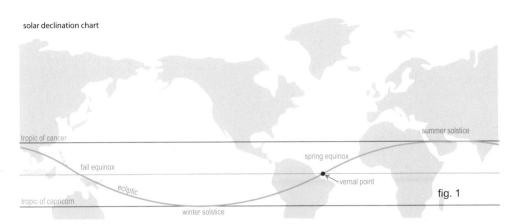

solar declination chart

tropic of cancer

summer solstice

spring equinox

fall equinox

equator

vernal point

ecliptic

tropic of capricorn

fig. 1

winter solstice

and below the equator. The point at which the Sun crosses from the southern hemisphere into the northern hemisphere each year is the "vernal point."

Through projection techniques of geodetic equivalency, we can project the ecliptic onto specific geographical coordinates. It should be noted that these placements use what is known as the "sidereal" placement for the vernal point. This is different in longitude from the tropical system of locating the vernal point. Over many years it has been observed that the sidereal placement is more effective for studying climate phenomena than the tropical system is.

When the Sun crosses the vernal point, it can be located through sidereal projection in the longitude of the eastern coast of Greenland. The solar zenith point in the Northern Hemisphere in June is reached as the Sun is moving over the Middle East. The nadir of the solar cycle latitude puts the position of the Sun at the winter solstice in December over the West Coast of the United States.

In the amazing wisdom of the cosmos, all of the other planets also move in longitude and declination throughout the year following the path of the Sun in time. This means that a planet such as Mercury, when transiting the Pacific, will have a tendency to move south in its declination, just as the Sun does in its yearly cycle. A planet such as Mercury, while transiting the Atlantic, will move north in latitude and declination, just as the Sun does in the spring of the year.

While the influence of the Sun moving along this path results in the phenomena of the seasons, other slower planets moving along this path are harder to link to phenomena. However, it is interesting to track a faster-moving planet like Mercury along the path of its approach to the West Coast of the

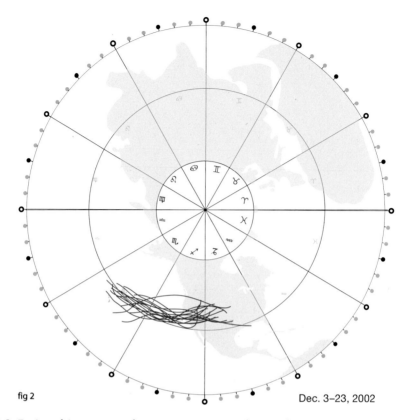

fig 2   Dec. 3–23, 2002

U.S. Doing this can reveal an amazing coincidence of polar jet stream oscillations related to the motions in arc of Mercury. For this study, the eclipse grid charts described earlier will be used.

In this image, we can see the daily paths of the polar jet stream in the Eastern Pacific between December 3 and 23, 2002. These paths are taken from daily newspaper records. We can see that the polar jet stream was very active in California, but not so active in the Pacific Northwest. This is a southerly path for the jet stream, a pattern that usually brings rain and storms to the south. At that time, Mercury was crossing the West Coast at its most southerly declination of 25° S latitude. This linking of the southerly placed jet and the southernmost transit in declination of Mercury might just be considered a coincidence that proves nothing, and that the southerly placed jet during the southerly declination transit of Mercury was purely random. Mercury was making a solo transit of the East Pacific at that time.

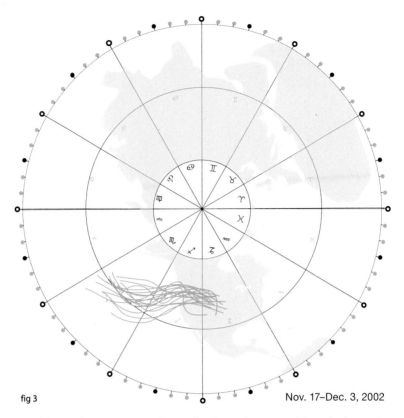

fig 3 Nov. 17–Dec. 3, 2002

To address this question, let us look at the period just before Mercury went into its southernmost declination. Figure 3 shows the daily jet stream paths in the Eastern Pacific from November 17 to December 3; this is the period for the two-and-a-half weeks just prior to the situation depicted in figure 2. We can see that the polar jet stream avoids the southerly latitudes and is centered in the Pacific Northwest for that time period. Even though the jet can sometimes be seen moving to a low latitude east of Hawaii, the prevailing motion is to the northeast as it approaches the coast. Mercury at this time was in the longitude of Hawaii, making its way eastward into the coast, and it was not at its most southerly declination. It was at a mid-point in its southerly run across the Pacific. The low-latitude tracks were in the longitude of the planet. However, as soon as Mercury reached 25° S declination on the third of December 2002, the polar jet stream shifted to the south (figure 2) and remained there during the entire maximum southerly

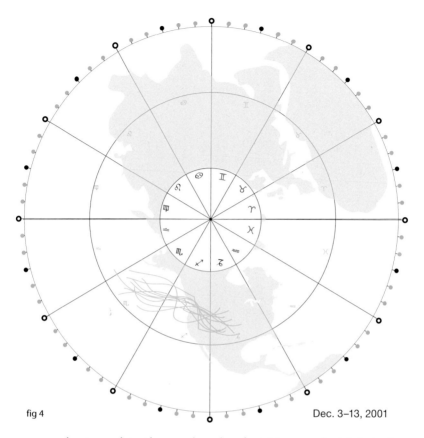

fig 4   Dec. 3–13, 2001

passage, and returned to the north only after Mercury had left its lowest point in transit.

This can be taken further by looking at 2001, the year before these events. Figure 4 shows the period from December 3 to 13, 2001. Mercury, moving in tandem with Venus, also at a low latitude, moved rapidly during 2001 as they approached the coast together. Mercury was moving about 1.5° each day in longitude; Venus was moving more slowly. As a result of the fast motion in longitude of Mercury, the period of its approach to the most southerly declination lasted only about ten days. Figure 4 depicts the polar jet stream motion in the Eastern Pacific during the time of the mid-latitude approach to maximum declination. Mercury in this chart is in the longitude of Hawaii. Even though the chart shows a bit more turbulence than the approach picture in figure 2, the same tendency toward low-latitude jet in the longitude of Hawaii turning north into the Pacific Northwest is present in this year.

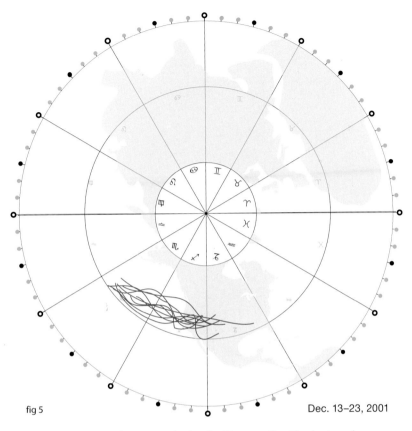

fig 5   Dec. 13–23, 2001

In figure 5, we see the jet tracks in the Eastern Pacific during the same year, 2001, at the time that Mercury was at its most southerly declination at 25° S latitude. Venus had been left behind. The signature of the turbulence of the rapid transit into the coast is evident, but the unmistakable signature of the southerly placed jet stream along the coast is overwhelmingly present, except for one day of tracking of the storm jet into the northwest.

In figure 6, we see the positions of the jet stream during the first half of November 2004. Mercury is again traveling solo. The period of Mercury's mid-latitude approach (blue) to lowest declination started on November 11, when the planet was near Hawaii. The period of its entrance into lowest declination (red) began November 17, when it was on the coast. In blue, we can see the now-familiar pattern of motion into the Pacific Northwest, when Mercury is near Hawaii approaching the farthest southern declination. The red lines depict the daily jet stream curves starting on November 16, the day of the

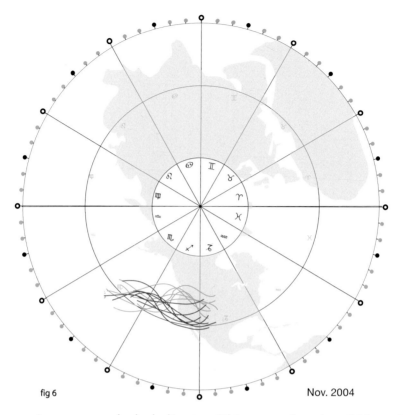

fig 6   Nov. 2004

entrance into most southerly declination. This pattern lasted until November 23. The only red curve that goes through the Pacific Northwest is the one for November 16, the day of the shift. Otherwise, the splitting of the jet stream by the Mercury entry to its most southerly declination is once again depicted precisely by the jet tracks from these two consecutive time periods.

Mercury is not the only planet to have an influence through declination. During December 1997, Mars was transiting the West Coast at its maximum southern declination. This influence was coincident with the formation of omega blocks (Ω) in January 1998 that brought flooding to southern California. The southern passage of Mars seemed to pull the jet stream to the south on the West Coast. This pervasive influence was coincident with the placement of low latitude low-pressure areas around the bottom of recurrent omega formations off the coast.

Figure 7 shows the eclipse grid pattern for flooding in Southern California during the winter of 2005. The pattern in 2005 was similar to the situation

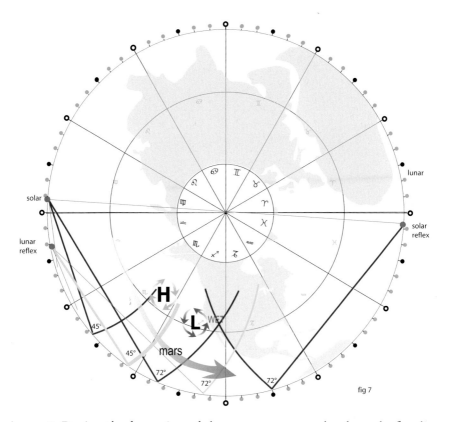

fig 7

in 1998. During the formation of the omega patterns that brought flooding
to southern California, Mars again was crossing the West Coast at a maximum southerly declination. This was again coincident with a strong southerly
positioning of the jet stream, and it was evident during most of the winter
of 2004 to 2005. Of note in this issue of declination is the fact that in 2005
Mars crossed the West Coast on February 21. The longitude for that crossing
is 130° W longitude (15° Sagittarius). It has been observed repeatedly that,
when a fast-moving planet is transiting the eastern Pacific, such a transit is
often accompanied by a southerly placed eastern Pacific jet stream bringing
wind and rain. It has also been observed many times that these turbulent phenomena subside when the transiting planet crosses 15° Sagittarius. The Mars
crossing was a dramatic example of this pattern. On February 21, as Mars
crossed the coast, the low that had been locked in place off the coast for seven
days slowly began to lumber to the east. By February 26, it was over Texas,
ending a remarkable set of storms for Southern California.

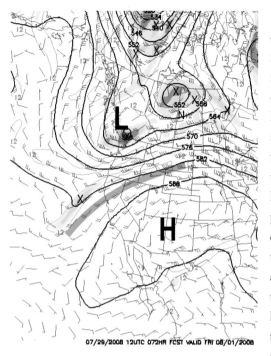

07/29/2008 12UTC 072HR FCST VALID FRI 08/01/2008

Finally, a baffling set of climate patterns in the summer of 2008 also had the markings of declination influences. Predictions of extreme heat on the West Coast for summer 2008 proved consistently wrong. The analogs for that summer showed strong heat arising whenever high pressure was present over the desert southwest. Repeatedly, the ridges would form there and heat would move toward the West Coast, only to be held back by a persistent trough formation off the coast of British Columbia. In the chart, we can see the ridge over the desert and the trough over the northwest, with the dividing line between them rendered in red.

This dividing line was problematic, since heat to the south was kept at bay by coolness to the north. This was not random but persisted through the summer. As a result, forecasts for heat in the wine country of coastal California never took place through July and August, even though there were many opportunities for ridges to build over the western portions of the desert.

In the analog year to 2008, heat was an issue when ridges built over the desert. In 2008, it seemed as though some force was allowing the troughs from the north to sag toward the south just enough to lock against the desert ridge, even when all of the influences on the eclipse grid shifted the patterns on the jet curves to high pressure. In analog years, that kind of pattern sent the temperatures soaring.

Then it was noticed that the position of Jupiter in 2008 put it into an approach to maximum southerly declination during June and July. The chart shows the condition of the eastern Pacific during the time of the approach by Jupiter to maximum southern declination. At the time, it was just off the West

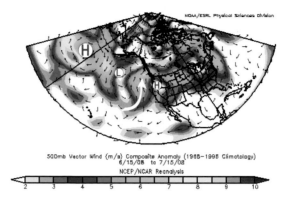

NOAA/ESRL Physical Sciences Division

500mb Vector Wind (m/s) Composite Anomaly (1968-1996 Climatology)
6/15/08 to 7/15/08
NCEP/NCAR Reanalysis

Coast. The variable in the National Weather Service chart is known as the "vector wind," the dominant flow in the upper atmosphere. The chart shows a consistent ridge over the western Gulf of Alaska, with a persistent moderate trough in the central Gulf of Alaska and a moderate ridge over northern California and Nevada. The circulation around the moderate trough kept the West Coast in a flow (white arrow in the figure above), where the marine layer off the coast was pressed into the coast for most of the month. This moderated the effects of the moderate ridge over northern California and Nevada.

The next chart shows the conditions in the eastern Pacific when Jupiter reached maximum southern declination in mid-July 2008. While the previous month-and-a-half had the dominant circulation from the south to the north in the middle of the Gulf of Alaska, the situation, once Jupiter had reached the bottom of its declination curve, was coincident with a strong shift to onshore flow coming off the same ridge over the western Gulf of Alaska. The bottom

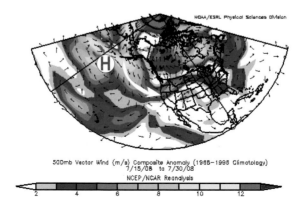

500mb Vector Wind (m/s) Composite Anomaly (1968-1996 Climatology)
7/15/08 to 7/30/08
NCEP/NCAR Reanalysis

of the onshore flow was in the same longitude as the position of Jupiter at maximum southern declination. The only real condition that shifted during these patterns was the onset of maximum southern declination by Jupiter just off the West Coast of the United States. Because of this influence, any ridging over the desert southwest was met by a persistent flow from the northwest. This prevented the desert heat from pushing up into the West Coast in July and early August in 2008.

CHAPTER 5

THE FORMATION OF ANALOG YEARS

Climate is the long-term patterning of short-term weather events. A weather event such as a storm is composed of similar weather forces, whether it is a large or small event. The difference is in the degree of severity of the forces. To study these forces is to study weather. However, not all winters are composed of the same forces. Winter one year can be radically different from winter the next year. The task of a climatologist is to understand the difference between a season in one year and that same season in another year. Climatologists use past weather patterns to predict future climate trends. To do this, they use analog years in which the weather patterns are studied and compared in detail to other years in which similar weather patterns emerged. Using these similar climate scenarios, a climatologist will look for signals that a coming season resembles the past analog patterns. Forecast accuracy better than 50:50 for these kinds of analogs is generally limited to a week into the future.

The eclipse grid model, which is the basis for this book, includes a fundamentally different kind of analog year to make predictions. The predictions based on these analogs have consistently provided a sixty-five to seventy percent accuracy rate a year in advance. The positions of the eclipse points in specific longitudes form the basis to formulate these analog years. The semi-annual shift of eclipse points creates decades-long rhythms that link years in statistically significant series. Sets of years that share common eclipse positions in longitude are the basis for climate analysis used in the following studies. The decade-long rhythms of eclipse positions have proved to be a reliable backdrop for climate research. In climatology, a recognized pattern in which similar events return in approximately ten-year intervals, is known as a "decadal return," or decadal influence, which are standard tools for the climatologist. In the eclipse grid model, the decadal influences are based on the approximately 9.3-year rotation of the lunar nodal point through half of the zodiac.

Every season includes weather patterns that dominate in different sections of the country. Warm air currents rising from the south and cold air currents descending from the north are the source of these patterns. Precipitation occurs where the cold and warm currents meet. During certain seasons, warm currents dominate in a specific area, and in other seasons cold currents will dominate. In the Midwest, for example, cold air from the north generally dominates during the winter, but warm air from the south generally dominates in the summer and late spring. These seasonal patterns are known to meteorologists as the *climatology* of the area. However, sometimes an area's climatology is not the pattern that unfolds.

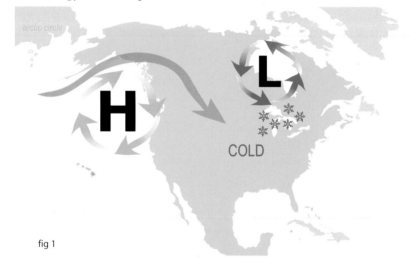

fig 1

For instance, sometimes in late spring very cold air dominates the Midwest (image above), creating late unseasonable snows that damage the tender corn seedlings. Before this can happen, the jet stream (as noted, a current of rapidly moving air in the upper atmosphere) needs to steer north into Canada over British Columbia. In the case of an atypical Midwest spring, such as the record May snowstorm in 1990, the agent that steered the jet stream north on the West Coast was the formation of a very strong high-pressure area, or blocking ridge, against the Pacific Northwest in the Gulf of Alaska. Blocking ridges restrict the flow of a jet stream that is traveling from west to east. In normal years, there is a strong high-pressure blocking ridge over Hawaii, but it does not usually push up far enough into the Gulf of Alaska to bring an abnormally cold trough down into the Midwest. In May 1990, the placement

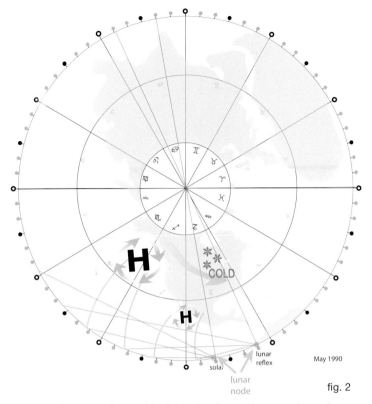

fig. 2

of a blocking ridge over the Gulf of Alaska formed a trough to the east of the ridge that was the path of abnormally cold air coming from the north into the Midwest. What would cause a ridge to push so far up into the Gulf of Alaska to bring an abnormally cold trough to the Midwest during summer?

The eclipse grid model tries to answer this by looking at the position of eclipses and the relationships between the eclipse points and other planets as a way to understand climate. As explained in earlier chapters, from each eclipse point, a set of jet curves is used to pinpoint areas of potential ridge formation. We can rely on the jet curves to be the favored areas where ridges will form during the six-month period between one eclipse and another. In a given season, the jet curves from the previous eclipse tend to be the places where unusual events connected to the placement of blocking ridges will most likely occur. In the chart (above), the lunar node is between the eclipse points. At the time of the cold wave in May 1990, the node was forming high-pressure aspects to both points. As a result, ridge formations dominated the jet curves projecting from the two points, especially over the Gulf of Alaska. The jet stream modeled itself on the ridge there, creating a record weather event for late spring.

However, if in a given season it is possible for the lunar node to stay in a sustained high-pressure aspect to the eclipse points, then the ridges would be persistent and a climate event would evolve from a weather pattern. This illustrates the practicality of using planetary motion as the basis for forming analogs. Furthermore, since the eclipse points travel in an east-to-west movement through an eighteen-and-a-half-year period, the position of potential blocking ridges travels westward along with them in predictable cadences. This rhythmic motion of the eclipse points and the jet curves associated with them provide insight into the decadal time frames and the duration of climatic patterns of the past, allowing analogs to be built that can support the formation of reliable predictions of the future.

Case studies of analog years when the eclipse points were in similar positions are the basis for the later climate studies in this book. In normal climatology, case studies are made on the basis of years during which the weather patterns are similar. In the eclipse grid model, similar analog years may have very different climate patterns. This may sound confusing, but analog years in the eclipse grid model are based on the common placement of the eclipse points and jet curves. Whether a jet curve is the site of a ridge or a trough in a given year depends on the motion of the planets forming aspect to the jet curves that year. The varying influences of these different positions of the planets approaching or moving away from the eclipse points in a given season are the basis for the predictions.

To show how this might work, it is helpful to compare the conditions in January 2005 to the conditions in January 2006 in the Midwest. The first chart (opposite) shows the eclipse grid for the winter 2005. There was a bitter cold pattern in the Midwest in January 2005. A strong ridge over the West Coast kept the jet stream (blue arrow) tracking north into western Canada out of the Gulf of Alaska. From there it descended into the Midwest, bringing blizzards and extreme cold during January and February. This pattern ended with the eclipse in April. The eclipse point shifted the jet curves to the west. In the climatology of the southeast, when the 45° jet curve line shifted onto the East Coast (red arrow), the ridge that was off the coast over Bermuda

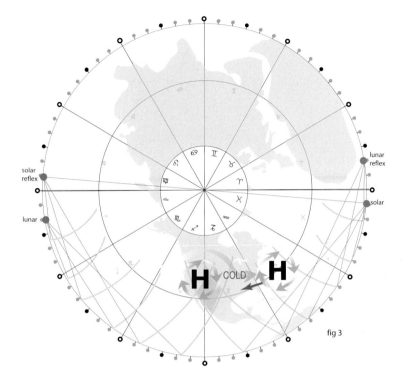

fig 3

moved onto the continent. While the ridge was off the continent in the early winter, Chicagoans were beaten down by cold and snow. In April, after the eclipse on the 8th, the ridge shifted along with the new jet curve position, and a high surged north into the Great Lakes, bringing hot and dry weather to the Midwest.

In January 2006, the shift of the eclipse grid during the intervening year placed both 45° jet curves straddling the eastern sections of the continent, while the shift simultaneously moved the pair of 72° jet curves well to the west, actually straddling the West Coast. The westward motion of the eclipse points that year had created a very different condition for the action of blocking high pressure. Coincident with this profound shift of the eclipse grid, January 2006 was one of the warmest on record for the continental U.S., so much so that Chicagoans were playing golf in shorts in winter. The development of high-pressure values on the new configuration placed a blocking ridge over the East Coast that drove the continental jet stream to the north. The zonal (horizontal) flow that resulted from the eastern ridge simply

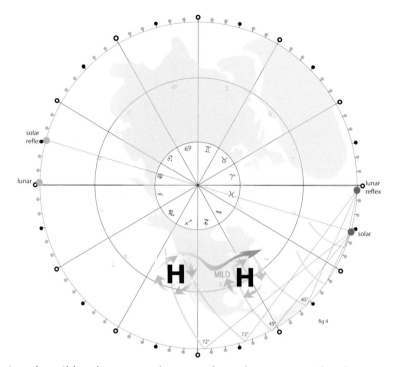

fig 4

continued a milder-than-normal pattern from the western ridge that was now halfway over the water of the eastern Gulf of Alaska. The jet stream moving around the top of the West Coast ridge dropped south through the mountains, but then ran into the south-to-north circulation from the western side of the East Coast ridge. The result was a radically different winter scenario for the two subsequent years.

This case study illustrates the unique value of the eclipse grid migrations as a method for building analogs to study climate patterns. We can take this much further. The following sequences show the transit of the eclipse points across the continental U.S. from February 1989 through July 1991. The configuration of the eclipse grid presents a harmonic spacing arrangement that is synchronous with the standard intervals of the loops of the jet stream. These loops, known as "Rossby waves," are spaced in an oscillating wave-train formation across the hemisphere. Predicting the placement of the Rossby waves (or "long waves," as they are called by climatologists) is important for the correct placement of the blocking ridges that steer the jet stream across the continent.

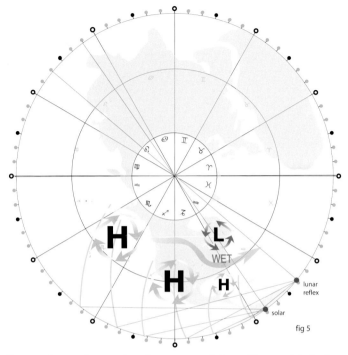

fig 5

The figure above shows the placement of the eclipse grid during the winter and spring of 1989. We see that, if a planet were to form a high-pressure aspect to the pair of points, a string of highs would form on the eclipse grid at specific intervals. The blue arrow follows the jet stream steering around the ridges. The high-latitude 72° jet curves above the western Gulf of Alaska support a ridge in that high-latitude area. The two 45° jet curves over the West Coast support a ridge at mid-latitude, and the two 22° jet curves support a ridge in the Gulf of Mexico at low latitude. This kind of flow patterning for the Rossby waves is known as a "cascade" in the language of the eclipse grid model. The jet tends to cascade down the ridges that are progressively lower in latitude. This is a commonly observed phenomenon.

We can see that a cascade of the long waves allows a center of low pressure to form in the longitude of the lowest latitude ridge. In this instance, the low is over Hudson Bay. This pattern is the exact 500mb ensemble for March, April, and May 1989. The Hudson Bay low was intense because the Eastern Seaboard was the site of anomalous rains and flooding in the early spring of 1989.

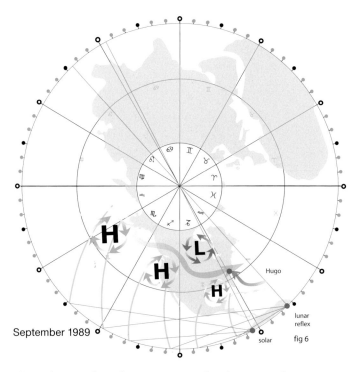

This chart depicts the eclipse positions for the next eclipse in August 1989. The eclipses have shifted the whole grid to the west. This has in turn shifted the ridge patterning to the west. The cascade pattern is still active, but the flow is farther to the west. The cascade has allowed a strong low to form in the Hudson Bay–Great Lakes area. The ridge on the 22° jet curves has blocked off the Gulf of Mexico from the incursion of any hurricanes transiting the Atlantic. In September, these kinds of patterns provide good opportunities for hurricanes to make landfall along the East Coast. The low over the Great Lakes is the source of troughs that can provide the steering mechanisms for storms transiting the western Atlantic. A cool flow off the Great Lakes meets the warm moist flow from the Bahamas, and large storms are the result. These conditions combined to support the momentous landfall of Hurricane Hugo in South Carolina in the third week of September 1989.

The next chart (opposite) shows the westward migration of the eclipse grid after the January 1990 eclipses. The cascade centers on the Midwest as a strong low formed over the Canadian Plains bringing flooding conditions to the Mississippi Valley in April and May of that year. The storm patterning

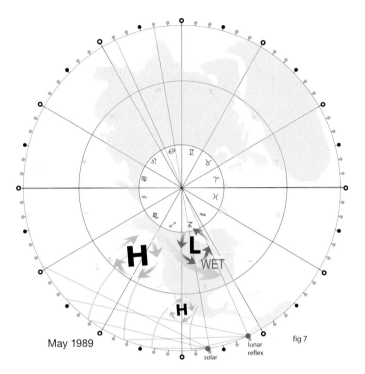

May 1989   lunar   fig 7
reflex
solar

follows the westward migration of the eclipse grid, as the cascade pattern in the Rossby waves slips farther to the west with each eclipse.

Figure 8 (page 154) illustrates the position of the next eclipse points and the Rossby patterns that led to flooding in the High Plains in the late summer of 1990. A monsoon condition arose because the mirror of the cascade pattern became active in the continental U.S. with the new eclipses. We can see that a mirror of the cascade is created when the projection of the jet curves continues east out over the Atlantic. The jet stream has a tendency to step up in longitude to hit the crests of the ridge patterns in a form, called a "berm" in the language of the eclipse grid model. The cascade and berm patterns create a symmetry in air masses that is far from being just theoretical. A summer-long study of symmetrical air-mass relationships between the Midwest and the U.K. during summer 2007 will be presented at the conclusion of this book.

In figure 8, the position of the low-pressure area on the continent was to the far west of Hudson Bay. That position steered the jet out of the Polar Regions for a good part of the summer and provided unusual and persistent thunderstorm activity on the High Plains as the monsoon moisture (red

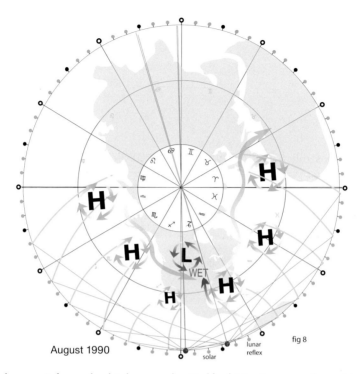

August 1990

solar   lunar reflex

fig 8

vertical arrow) from the high over the Gulf of Mexico met the cold air cascading down from the Gulf of Alaska. As the new eclipse pattern unfolded for the next eclipse, the berm pattern began to exert the dominant air-mass relationships for the continent. Obviously, not all values on the eclipse points were high-pressure values during the period of these examples.

This section is intended to show how the formation of analog-year patterns can be used to observe extended-period weather anomalies and place them into a rational and workable modeling procedure based on the westward migration of the eclipse points in decadal rhythms.

CHAPTER 6

PACIFIC NORTH AMERICA PATTERN

It is now possible to begin illustrating case studies to support the more theoretical or anecdotal examples presented so far. To do this, it will be instructive to look at a prominent climate pattern for North America. On Christmas 2004, Texas experienced a record snow. Five days later, the temperatures in Texas were in the seventies and eighties. In 2004, in the middle of a winter heat wave in Texas, the neighbor to the north, Colorado, experienced heavy snow. In late December 2004, people in the Carolinas were skidding across the roadways on black ice, and a few days later, they were basking in spring-like conditions. The culprit behind these events was a shifting seesaw system known as the "Pacific North America (PNA) pattern."

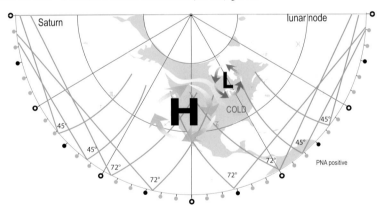

The first image shows the conditions in mid-December, when a strong high over the West Coast pushed the jet stream up into Alaska, causing temperatures to plummet in the eastern half of the continental U.S. This pattern occurs when the east Pacific high-pressure area comes close to the coast. In this position, the jet stream cannot drop along the West Coast, but is pushed up over the Canadian Rockies. In mid-December 2004, this pattern was supported by a series of planetary aspects that produced strong high-pressure against the coast (blue 72° jet curves). Saturn in the west and the lunar node in the east both put high-pressure values on the respective eclipse points, shifting the whole grid to high pressure. A robust blocking ridge formed over the area of the eclipse grid where the intersection of both sets of 72° jet curves was positioned (yellow diamond). This high then settled over the inter-mountain area and effectively blocked the jet stream into western Canada, pushing it far to the north. The result was abnormal cold in the Midwest that went down into the Gulf Coast, creating snowstorms in Texas and ice in the Carolinas.

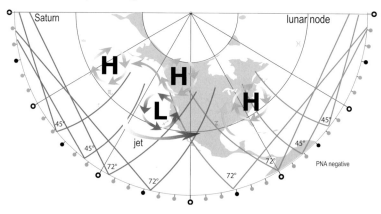

This image shows the position of the jet stream after a Saturn motion-in-arc during the last week of December 2004 shifted the western portion of the grid to low-pressure values over the eastern Pacific (45° jet curves in red). The pattern illustrated here is the negative PNA pattern. The blocking high that was over the western mountains shifted north and moved into the western Gulf of Alaska. This shift allowed the polar jet stream to drop to the south along the West Coast, bringing a cold low down through California and on into the Denver area. As seen in the chart, the shift of the blocking ridge in the west has also allowed a high to build over the Gulf Coast on the jet curves linked to the node (blue). The circulation around this broad high brought warm moist Gulf of Mexico air into the continent (green arrow). This was the feature that supported shirtsleeve temperatures along the Gulf Coast and Texas, where it had been snowing a few days earlier. The shift of the high to the north and west in the eastern Pacific was coincident with Saturn motion-in-arc event that changed the influence in the Pacific from high pressure to low pressure (red Pacific jet curves). This shift allowed the polar jet to drop to the south more easily on the West Coast.

These sequences clearly show two levels of activity used to analyze future patterns. The two levels involve planetary configurations in different ways. The first level is the fundamental placement of the eclipse points in longitude. These positions project the eclipse grid onto the Earth in specific longitudes. The most sensitive structure in the eclipse grid is the disturbance diamond created by the crossing of the intersection of both sets of 72° jet curves. The first chart showed this in yellow.

The second level of activity in the charts is the angular aspects of the planets in relation to the eclipse points. The angular aspects of Saturn and the lunar node were active in shifting these values for the opposite PNA patterns. These shifts were in line with the values given in the article about planetary approach and retreat.

HISTORIC CITRUS FREEZES

The patterns that give rise to freezes in citrus areas of the United States have a pronounced periodicity when placed in the context of eclipse rhythms. To see how the eclipse rhythms influence climate patterns in general, we need to understand the time dynamics of the eclipse grid. The following case studies in the next few sections all revolve around the placement of the eclipse grid in specific longitudes at specific times. They are intended to show that eclipse influences can provide a reliable research tool for studying climate periodicities. This is because the natural cycles of 9.3 and 18.6 years for the movement of the lunar node through the zodiac are in sync with many climate patterns that occur in decade-long (decadal) rhythms or two-decade (inter-decadal) rhythms.

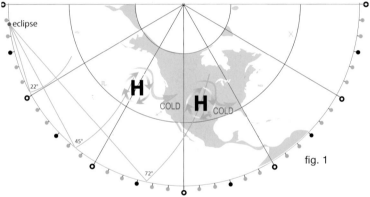

fig. 1

The first image illustrates the formation of two jet curves, one at 45° of arc and the other at 72° of arc using an eclipse point as the center of the generating area. In a given winter, the polar jet tends to track along these curves when forming stable, high-pressure ridges of air. Where the jet rounds across the top of a block, cold air is brought down the eastern side from the north to the south. Citrus freezes arise when the blocks are situated in such a way that the polar jet can drop the cold into California or Florida.

The position of the eclipse points is a potent variable in the formation of unseasonable frosts. In the chart above, the eclipse point in the western Pacific is projecting jet curves into the central Gulf of Alaska and into the Rockies. Cold is descending the eastern side of the block in the Gulf of Alaska. This is the kind of pattern that is usually involved in citrus freezes in California.

Major California citrus freezes

- 1913/October eclipse point western Pacific:
  January freeze southern California
- 1974/75/Nov 74: eclipse points eastern Pacific
- 1990/Dec 21 freeze: eclipse point eastern Pacific.
- 1998/Dec freeze: eclipse point eastern Pacific.

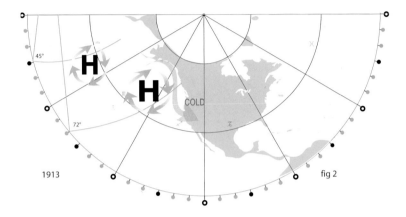

In California, the strongest citrus freezes have occurred when the eclipse points were in either the eastern or western Pacific. These positions place the eclipse grid in such a way that ridges can form in the central Gulf of Alaska. In January 1913, the granddaddy of citrus freezes almost wiped out the fledgling citrus industry on the West Coast just as it was getting started. The eclipse points from the September eclipse were in the western Pacific. The 45° and 72° jet curves from the eastern point are illustrated in the image above. The blue arrow shows the cold pouring down out of Alaska, following the ridge centered on the 72° jet curve. That year, ridge formation in the central Gulf of Alaska allowed the storm jet to enter southern California from Alaska, bringing sustained cold during the winter.

In December 1975, another citrus-freeze year, the eclipse points were in the mid-Pacific. The grid of jet curves from the easternmost point is illustrated in the next image (following page). In that year, the cold drove south along the eastern side of a ridge that formed on the 45° jet curve from the eclipse point with the same result.

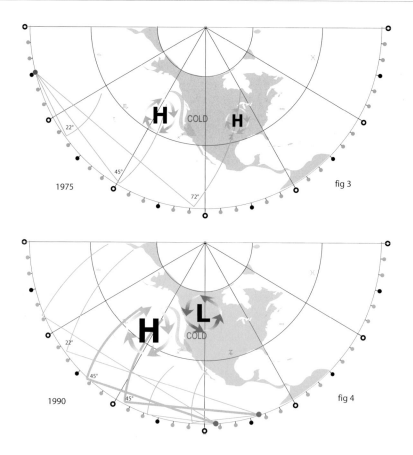

In the fall of 1990, the eclipse points were in the far eastern Pacific (image immediately above). This created another jet curve pattern in which a ridge on the 45° curves guided a looping jet stream trough down the West Coast from British Columbia on the downstream side of a blocking ridge. As a result, in December 1990 a strong citrus freeze unfolded in southern California along the downstream side of the ridge on the two 45° jet curves. The freeze started December 21, 1990, and extended for two weeks into January 1991.

The next citrus freeze began December 20, 1998, and lasted for a week, killing many citrus trees in California. The eclipse position was once again over Central America, putting the 45° jet curve parallel to the West Coast of North America. The next image (opposite) shows this pattern. This time, however, the key players were the two 72° jet curves from the eclipse points.

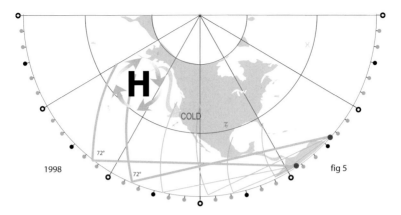

1998   fig 5

These high-latitude features put the ridge formation at high latitude in the western and central Gulf of Alaska. The high pressure from the two 72° jet curves pushed up into Alaska, and strong cold descended into California. The placement of the jet curves in the Gulf of Alaska determines the placement of the polar jet stream. In this instance, the offshore placement of the jet curves pulled the ridge offshore, allowing the cold to descend.

Of course, there were many other factors in these freezes. This short section is meant only to introduce the idea that the placement of jet curves in any particular year often coincides with the placement of the blocking ridges that form the jet stream motions during that period. It needs to be said that the planetary aspects that influence the eclipse points must strongly produce high pressure in order for the ridges to create a sustained block in winter that will bring down the cold. That being said, the placement of the eclipse grid could give general insights into the time dynamics of recurring climate phenomena. The general situation can, however, be analyzed in detail when harmonics of the Pythagorean monochord are integrated into the grid.

Florida citrus freezes

We can now use the ideas in the previous section about California citrus freezes to illustrate citrus freezes on the opposite side of the country. Instead of doing single years, however, we can group years in which the eclipse positions placed a jet curve in a local where it could influence ridge formation. This modeling approach allows a survey in which we can observe decade-long influences.

In the historic accounts of citrus freezes in Florida, certain dates stand out. The most intense period of freeze activity in the last few decades in Florida was during the 1980s. During that time there were five years in which significant freezing temperatures altered the citrus crop to the point of having economic impact. The years of the killing freezes were:

- January 12–13, 1981
- December 24–25, 1983
- January 21–22, 1985
- December 23–24, 1989

These were four killing freezes in a single decade. This pattern dealt a heavy blow to the Florida orange-juice industry. From the point of view of the eclipse grid model, a curious shifting of eclipse points took place during that time that sheds light onto this unusual weather pattern.

In 1981, the position of the eclipse points for the citrus freeze in January was over the western Atlantic (top image opposite). They were the points from the summer eclipse. This position projected a strong ridge on the 45° jet curves over western Canada. These points were aspected by Neptune and Uranus to strong high-pressure values for most of January. These high-pressure aspects were coincident with a strong high caught in western Canada in January. The strong high pushed the jet stream (arrow) to the north and brought cold down into the eastern parts of the continental U.S. This pattern strongly supported a freeze in the southeast. We can also see that two lows formed on either side of the western ridge. This is part of the general circulation patterns for North America. The low over Hudson Bay is entrained in the lee of the ridge over the Canadian Rockies.

In December 1983 (lower image opposite), the eclipse points had shifted to the eastern Pacific. In the time between the eclipses, Jupiter formed a strong

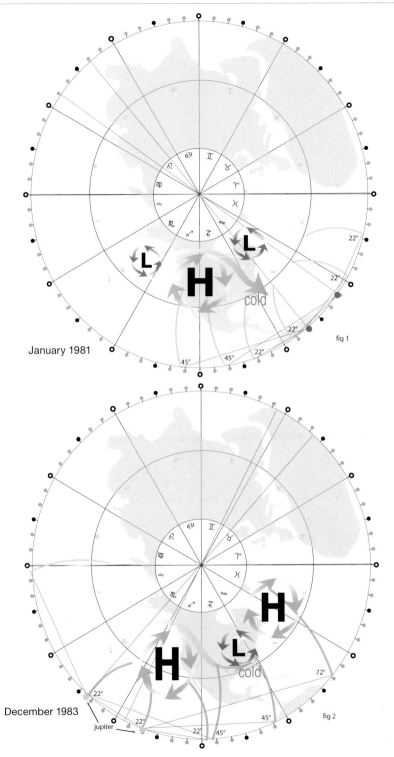

January 1981

fig 1

December 1983

jupiter

fig 2

high-pressure aspect to both sets of eclipse points (blue). The jet curves in this chart are oddly spaced, because the shift of the eclipse points skewed the spacing between eclipses. The jet curves in all areas of the chart points (thick blue curves) became the sites of strong high-pressure air masses. The two easternmost curves were off the continent. They supported a strong Greenland block pattern that pushed the continental jet to the south and into the High Plains. The 22° jet curves were focused over the eastern Pacific. That area was the site of a ridge that pushed the jet north along the West Coast. A trough was caught in the lee of the ridge and settled just south of Hudson Bay. The standard climatic polarity of an air mass of one sign (high pressure) in the Gulf of Alaska was linked to an air mass of the polar opposite sign (low pressure) over Hudson Bay created the climatic anomaly. The jet stream caught the flow between these two dominating air masses and drove strong cold into the southeast creating a very damaging freeze in Florida.

In 1985, the November 1984 eclipses had shifted the points to the central Pacific. In January 1985 (upper image opposite), a Jupiter aspect put a strong signal for what is called in this model a "lee trough pattern" onto the eastern eclipse point. By projection, the area between the 45° and 72° jet curves from that point was the site of a strong continental ridge. Just to the east of the ridge, a lee trough formed over the northeastern U.S. The lee trough signal is a geometric relationship that often is active in the formation of strong trough formations that are formed just in the immediate downstream position to a strong ridge. In this case, the ridge over Hudson Bay was dynamically linked to a strong lee trough over the Northeast. This pattern drove considerable cold down the east coast and into Florida, further damaging an already-weakened citrus industry.

A final blow to the citrus industry in Florida occurred during December 1989 (lower image opposite). A strong set of values from Neptune created a perfect condition for the record Pacific North American pattern when a high-pressure surge from Neptune on both 45° jet curves over the west pushed a strong ridge to the north. The downside of the ridge was a balancing activity in the form of a monumental lee trough that formed from the Mississippi Valley eastward, bringing cold air down from Canada into the southeast.

Again, this study is a simplification of the actual research protocols used to form the model. In practice, the position of the jet curves is not the only

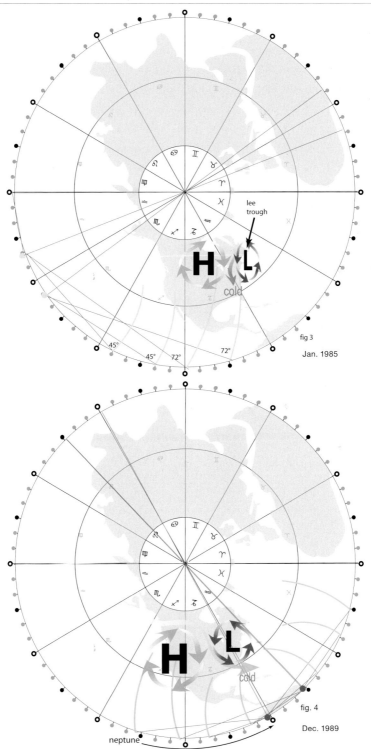

fig 3

Jan. 1985

fig. 4

Dec. 1989

influence in the patterns resulting in these freezes. Equally important in these patterns is that, when these curves were in these positions, strong high-pressure values influenced them from juxtapositions of planets moving into strong high-pressure aspects to the eclipse points. If this had not been the case, then each time the eclipse points reached these positions, there would be a citrus freeze. Since this is not the case, we can only marvel that the placement of planets in significant aspects was coincident with the required position of the jet curves in decadal rhythms culminating in the great onslaught of cold into the southeast during the 1980s.

The remarkable patterning in these relationships involving the position of the eclipse points and their rhythmic sequences of projected jet curves provides a useful way to study unusual climate patterns that repeat in specific places in decadal sequences.

THREE DENVER BLIZZARDS

Using the dynamics of the eclipse grid as a research tool frequently provides insights into how a recurring climate pattern in a given area can result from very different atmospheric structures. A good case in point is the climate pattern that results in the formation of blizzard conditions in the Denver area. This pattern is a complex of trough activity from the Gulf of Alaska meeting wet circulation around a ridge in the Gulf of Mexico. As we shall see, there are a number of different scenarios that can result in this interaction.

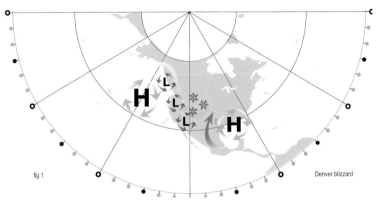

fig 1   Denver blizzard

The classic climatic pattern for the emergence of a blizzard in Denver involves the sudden movement of a cold air mass from British Columbia southward into the Basin and Range area between Denver and California. The upstream stimulus for the cold air to drop from a high latitude is often a high-pressure area situated in the central Gulf of Alaska that pushes the jet stream up into Alaska. There the cold air forms the cold front that eventually drops south along the West Coast. A low-pressure area usually accompanies the cold front. As the low moves into southerly latitudes, its counterclockwise circulation picks up moisture from the back end of a clockwise rotating ridge located over the Gulf of Mexico. The timing of this must be such that the low encounters the ridge influence just as it turns eastward at the bottom of its southward swing through the inter-mountain west. The cold low-pressure area draws warm moist air from the ridge into the leading edge of the circulation. The warm moist air streams up from Texas and into Colorado, dumping wet snow mixed with rain that can accumulate rapidly in the mixing conditions of the cold front meeting the warm monsoon. Snow forms easily and falls in abundance, bringing remarkable accumulations in a short time. These types of storms often make a dismal water year a success instead of a disaster. It would therefore be good to know in advance if there was an increased potential for these types of storms to form in a given water year.

In conventional climatology, the Denver blizzard pattern is linked most often to El Niño years, when the dominance of a high-latitude high-pressure area over the Gulf of Alaska splits the Pacific jet stream into two currents. Warmth from the El Niño in the eastern Pacific supports the formation of strong high pressure in the Gulf of Alaska. When this high forms, the storm jet divides into two currents. One rides up and over the block into Alaska, and the other drops south through southern California and the Desert Southwest. The split storm jet recombines over the southern part of the Colorado Plateau. The counterclockwise circulation of the storm draws the moisture from the Gulf of Mexico and delivers it to the High Plains. The record snowfall of the blizzard is the result. The big question is: Why in some years does this pattern unfold and then, for the longest time, there is a drought pattern in the High Plains? If the blizzards occurred only during El Niño years, it would be simple to predict that, when an El Niño event is underway, a blizzard season is probable. The catch, however, is that the pattern for Denver blizzards does not occur only during El Niño years.

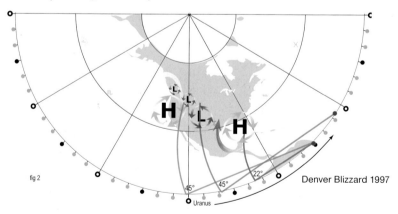

fig 2   Denver Blizzard 1997

The image above shows the eclipse grid and the accompanying jet curves for 1997. This was a major El Niño year, and the eastern Pacific was very warm during that October. The El Niño split-jet pattern was very strong in the fall, since the SSTs had been rising since the beginning of the year and continued through the summer, and were now peaking in the eastern Pacific at almost 2.5° above normal. This is a very strong warming. We can see from the eclipse grid (opposite) that the placement of the eclipse points is in the western Atlantic.

In 1997, the two 45° jet curves from the eclipse points over the western Atlantic were stimulated in late October. Uranus, approaching the eclipse points from the west (curved black arrow), moved in arc on October 19 to create a value of low pressure on the 45° jet curve over the Rockies (red), and a high-pressure signal on the 22° and 45° jet curves from the other point (blue). This placement put a high over the Great Basin that pushed up into Alaska. This allowed a weak low to drop down on the eastern side of the high. The cold low moved south into the Denver area along the low-pressure 45° jet curve (red). This lee trough pattern delivered a weak but cold trough to the Rockies that came down the West Coast along the eastern side of the surging high. The low then turned eastward toward the Rockies when it reached a southerly latitude. As it approached the low-pressure (trough-forming) 45° jet curve over the High Plains, it suddenly exploded into a major storm that vigorously drew warm and moist tropical air into its circulation. A ridge on the 22° jet curve that was over the eastern Gulf of Mexico provided the warm monsoonal flow that supported the explosive nature of this storm.

The image on the following page shows the eclipse positions and the resulting jet curves for the second week of April 2001. In this situation, the eclipse points are due south of the West Coast. Besides the 45° jet curves projected from the eclipse points, there is another set of jet curves that can be useful in the eclipse grid model—jet curves at 22° from the eclipse points. In April 2001, the 22° jet curves were placed in a similar longitudinal position to the 45° jet curves in 1997. Since the eclipse points were due south of the West Coast, there was also a set of 22° jet curves to the west of the eclipse points in the center of the Gulf of Alaska. Starting April 8, the approach of Mars toward the eclipse points coincided with a strong surge of high pressure on the two sets of 22° jet curves. This was coincident with high-pressure areas intensifying in the central Gulf of Alaska and over West Texas. Over the ocean, high pressure surged out of the middle of the eastern Pacific and up into the center of the Gulf of Alaska. Over the continent, high pressure

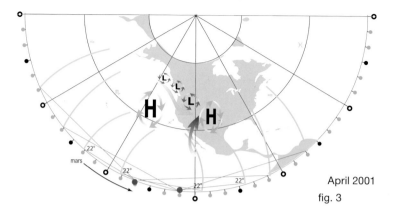

April 2001

fig. 3

pushed up over the Rockies and the High Plains. In between these two surging high-pressure areas, a weak low-pressure cell started to drift south along the West Coast. Cold air accompanied the low as it drifted across California into the Great Basin on the 10th. On April 11, Mars moved in arc to low-pressure values on all of the 22° jet curves. On that day, the transiting low consolidated along the now low-pressure jet curves over the western states. With this shift, the strengthening low dropped south and met a stream of moisture coming from the Gulf of California, which had been organized by the circulation of the high over West Texas. The weak but cold storm then turned into a Denver blizzard. In this sequence, it was not really a strong or significant El Niño that instigated the blizzard. The trough was formed in the weak area between the sets of 22° jet curves.

The image opposite shows the eclipse lines of the Denver blizzard of March 2003. That year, there was a strong La Niña cooling event taking place during the time of the blizzard, so the standard El Niño support for a southerly placement of the Pacific storm jet was not available. On March 15, a strong low-pressure area formed off the southern coast of Alaska. A continental high in the Pacific Northwest blocked this low from transiting through the north. Mars was the only strongly transiting planet at the time. It had passed the eclipse points in late January and was now crossing the coastline on the West Coast. When a planet crosses the coastline, a strong turbulence is often the result, since the planet crossing the coast is often moving at a maximum southerly declination (see chapter 4, third section). Sagittarius is the constellation that straddles the West Coast. It is the most southerly of the

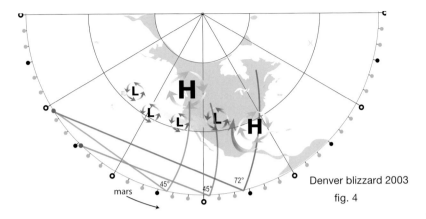

Denver blizzard 2003

fig. 4

constellations, and a planet moving though it is often at an extreme southerly declination. The passage of a planet at an extreme southerly declination is often accompanied by the southerly drooping of the storm jet.

In this case the storm jet dropped to the south on the West Coast as Mars transited the coast. This transit was simultaneous with the formation of a strong low-pressure value on 45° jet curve over the Rockies (red). A strong but short-lived low-pressure area on the 45° jet curve served as the seed for a Denver blizzard. In this sequence, the low-pressure area flowed beneath the blocking high over Washington State and then drifted east at a southerly latitude.

In conclusion, the eclipse grid-modeling technique can provide insights into climate patterns that occur in odd, or non-rhythmic, sequences.

WINTER IN 2005—A YEAR OF UNUSUAL WEATHER

The citizens of northeastern U.S. experienced a record number of snow events during the winter of 2005, when an archetypal pattern in the eclipse grid was coincident with nineteen snow events. On the West Coast, 2005 was also a banner year for rain and snow in southern California. This section places both of these unusual events in the context of archetypal North American climate patterns that were stimulated by precise placements of jet curves across the eclipse grid in North America in the winter of 2005.

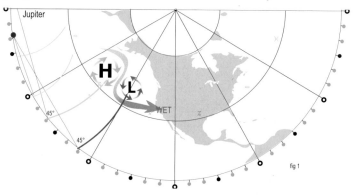

Figure 1 shows the jet curves for the most western portion of the most common pattern in the winter of 2005. The 45° jet curves over the Gulf of Alaska were aspected by Jupiter to a split between high pressure to the west near Hawaii (blue curve) and low pressure on the eastern point, projecting to the central Gulf of Alaska (red curved line). This pattern brought the jet stream in the eastern Pacific just to the east of Hawaii. Cold from the western Gulf of Alaska dropped south and then curved into the West Coast (blue/red arrow). The alternation of high pressure and low pressure on the successive jet curves is the situation that gave the winter its particular dynamic for the West Coast. The position of the high- and low-pressure air masses is a general placement for that winter. Of course, the air masses shifted during the winter, but a mean position in the eastern Pacific for the winter months is shown.

Figure 1 shows only two 45° jet curves (figure 5, page 178, depicts the complete set for the winter). During that winter, the shifting values on the jet curves was greatly influenced by the position of Jupiter over the western Pacific in the locale of the eclipse points (a red dot for one point, the other point is not shown). The activity of Jupiter near the eclipse points during this winter kept this pattern strongly in place, forming the ridge-and-trough sequence seen in the chart. The result of this ridge-and-trough pattern was unusual wetness in southern California.

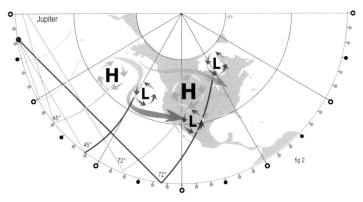

Figure 2 continues the pattern of high pressure on the western curve and low pressure on the eastern curve shown on the 45° jet curves over the Gulf of Alaska. Added to these curves are two 72° jet curves over the intermountain area, extending into the Hudson Bay area. These curves link to the same eclipse points over the western Pacific as the two 45° jet curves in figure 1. The western 72° jet curve is aspected to high pressure by Jupiter (blue curved line). This puts a high-pressure area across the intermountain west. The eastern 72° jet curve from the western pair is linked to the point that is aspected to low pressure by Jupiter (red curved line). This jet curve crosses the southern tip of the Baja peninsula and continues northeast toward Hudson Bay. This type of pattern most often resulted in a split jet stream that winter. The two arrows in fig. 2 show one part of the jet moving to the south, below a high over the western states, and another weaker leg of the polar jet moving across the top of the ridge, to descend into the Great Lakes area and flow along the southern border of a low stationed over Hudson Bay on the eastern 72° jet curve. This pattern set the stage for the unusual weather on the continent. To get the complete picture, however, we need to shift to a whole-hemisphere map so that all of the jet curves can be seen.

Figure 3 (next page) shows how the two 72° jet curves from the western pair of eclipse points illustrated in figure 2 are crossed by the 72° jet curves from the eastern pair of points over western Africa. In the winter of 2005, the

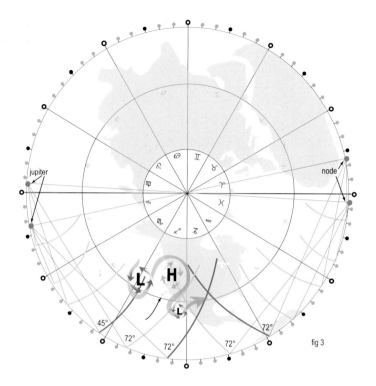

fig 3

crossing forms a diamond (yellow) over the Southwest. For most of the winter, the motion of the lunar node influenced the two eastern eclipse points and, by projection, the jet curves associated with them. The most prominent pattern was that the westernmost 72° jet curve from the eastern pair of points (blue curve from Mexico to Vancouver Island) was aspected to high pressure for long periods of time. The easternmost 72° jet curve from the eastern pair of points was aspected to low pressure for most of the time (red curve from Oklahoma to Montana). This pattern was a persistent peculiarity of this particular season.

All four 72° jet curves taken together form an eclipse diamond in their crossing (yellow diamond). In figure 3, we see that these relationships put the two 72° jet curves on the western side of the eclipse diamond under high pressure and the two 72° jet curves on the eastern side of the eclipse diamond under low pressure. This is very significant in this chart and it was a very persistent pattern during that winter. It put a strong block over the West and a trough formation over the Southwest and southern High Plains. That pattern allowed storms from Southern California to pass over the mountains and form again over the middle of the continent. From there, the usual track was across the Gulf coast and then up into New England. It also allowed the storm track in the west to form an unusual looping jet that climatologists call an "omega pattern."

Figure 3 shows an omega (Ω) pattern over the western states during this winter. The eastern 45° jet curve over the Gulf of Alaska was the site of a persistent low in the center of the Gulf of Alaska. The low-latitude low then encountered the high-pressure ridge situated in the Pacific Northwest, near the two 72° jet curves on the western side of the eclipse diamond. One leg of the two crossed 72° jet curves extended into the tropical part of the East Pacific. The ridge associated with this jet curve often emerged from the eastern tropical Pacific and pushed northeast along the jet curve and across the coastal mountains (small black arrow). From there it snaked into the western inter-mountain areas to meet the northern leg of the other 72° jet curve that ran from Mexico into the PNW. With this surge of high pressure to the north, which linked the two 72° jet curves on the western side of the diamond, the low to the west over the middle of the Gulf of Alaska dug south. Another low off the coast of Southern California was situated on the tropical section of the 72° jet curve, which was under low-pressure influences (red). These two lows were locked in place by the long vertical blocking ridge along the western mountains. These relationships set up the omega pattern. In it, the jet stream was directed by the strong high fixed over the northwestern U.S. The jet stream surged to the north along, around, and over the high, and then plummeted south around the northeastern side of the block.

Starting in the west and looking at the complete pattern, the jet stream moved south just to the east of Hawaii along a low-pressure 45° jet curve. It then veered north and east into western Canada. From there, the high pressure steered the jet once more to the south along the West Coast. Running south, it turned east in tropical eastern Pacific. The whole motion formed a horseshoe, or omega. This omega was the western formation of a very common climatic pattern that unfolded during that winter. It is clear from the many daily charts at the 500mb level that a persistent high centered just to the northwest of the crossing point of the two high-pressure 72° jet curves over the west. This pattern brought much rain to the south and dryness to the PNW.

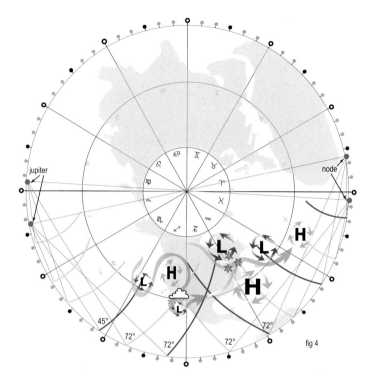

fig 4

In figure 4, the 45° jet curve lines for the eastern pair of eclipses are included in the chart. We can see that the westernmost 45° jet curve of the eastern pair is influenced to high-pressure values (blue) in the vicinity of Bermuda. This curve supported a low-latitude ridge off the East Coast in that area. The easternmost 45° jet curve (red, or low pressure) is placed with its northern tip toward Canada's Maritime Provinces. This placement supported the emergence of a low-pressure area linked to the Hudson Bay low, a winter feature of the northern hemisphere climate. A large "L" over Hudson Bay represents the Hudson Bay low. We see that the Hudson Bay low is at the tip of the more easterly of the two crossed 72° jet curves over the central U.S. A strong low here dominated the continental regime. This condition combined with the offshore low-pressure area south of the Maritimes to enhance storm activity in the northeast. The trough formation over the Northeast and the symbol for snow represents this pattern.

The 45° jet curve in the Bermuda area was the site of a persistent high-pressure ridge (blue). This high created a blocking pattern that prevented the storm jet across the continent from easily exiting North America. This high steered moisture from the Gulf Coast up into the cold coming down from the Hudson Bay low (green arrow). The result was a persistent trough formation over the northeast U.S., with a block to the south in the western Atlantic. These jet curve aspects were persistent owing to unusual motions of the lunar node aspecting the eastern eclipse points during the winter months. In response to the western Atlantic low-latitude ridge, the continental jet stream coming in from the southwest tracked across Texas and then across the Midwest, bringing much rain and cold. The low-latitude high in the western Atlantic then nudged the fronts into the northeast, and abundant New England snows were the result. The small green arrow moving up the coast toward New England represents this track. The source of the fronts that energized these storms was the counterclockwise circulation around the low over Hudson Bay.

To extend the implications of the potential for the eclipse grid as a tool for climate study, the whole set of jet curves can be placed in context for the winter of 2005. Starting in the west, a moderate high is on the 45° jet curve from the western pair of eclipse points. At the time, Jupiter formed a persistent high-pressure aspect to the western point over the mid-Pacific. The moderate ridge on the 45° jet curve linked, or "tele-connected" (as it is called in climatology), to the 72° jet curve from that same point. We see this in the blue horseshoe-shaped air mass over the eastern Pacific. This tele-connection caused the omega to form over the West Coast, with its two accompanying low-pressure areas. With the eclipse diamond divided into high-pressure values on the western two 72° jet curves and low-pressure values on the two 72° jet curves on the eastern side of the diamond, the tele-connections from the Pacific to the mid-Atlantic engendered several archetypal climate patterns.

An omega in the west is most often balanced by a strong Hudson Bay low in the east. This pattern is well documented. In this case, the low-pressure values on the eastern side of the eclipse diamond supported the archetypal pattern. The other archetypal climatic pattern was the tele-connection between the Bermuda high and the Azores high. The Bermuda high was influenced by high-pressure values generated by the lunar node near the eastern eclipse points over western Africa. On the chart, we see that the Bermuda high was situated on the 45° jet curve. It happened often in the winter of 2005 that, when this curve was aspected to high pressure, the ridge built

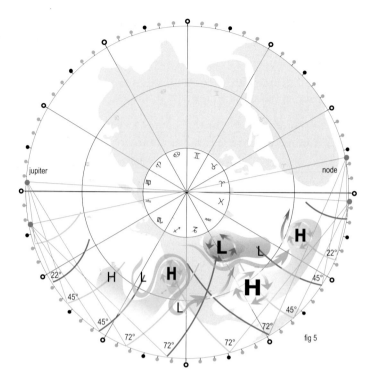

jupiter

node

fig 5

up in the western Atlantic at a low latitude and then drifted east to connect to the 22° jet curve over the central Atlantic. When the two connected, a strong ridge built up there over the Azores. This is an archetypal climate pattern for the Atlantic. In the chart above, the horseshoe shaped high over the Atlantic is the beginning of the blocking pattern. When that happened, the low-pressure 45° jet curve that pointed to the eastern tip of the Maritimes became the site of a strongly sagging low-pressure trough that was fed from the Hudson Bay low.

In response to the oscillation of the central Atlantic ridge and the western Atlantic ridge, the U.S. continental jet stream coming in from the southwest tracked across Texas and then across the Gulf Coast. The high in the western Atlantic then nudged the fronts into the Northeast, and abundant snows in New England were the result. The small arrow moving up the coast toward New England represents this track. The source of the fronts that energized these storms was the counterclockwise circulation around the low over Hudson Bay.

Taken together, the omega on the West Coast and the strong trough in the Hudson Bay area, kept in place by an active ridge formation in the mid-Atlantic, fit the prevailing pattern present in the relationship between Jupiter in the west and the lunar node in the east as they moved to create different aspects on the eclipse grid. Because of the periods of motion-in-arc, the rhythms of the lunar node and Jupiter are the most synchronous of all the planets. There are periods when these two influences can dominate a chart for months. They were effective in this winter scenario for the production of rain and snow. In a later chapter, we will meet the combination of Jupiter and the lunar node again when they team up to create a strongly polarized pattern of dryness and flood in the Midwest during the summer of 2007.

A dust storm approaches Stratford, Texas, 1935

CHAPTER 7

DROUGHT

The United Nations Development Program (UNDP) found that drought is the most important natural hazard in terms of human mortality. Seven of the ten most drought-prone countries in Africa are those that have also suffered most either from armed conflicts or from political instability, with more than fifty million people affected by severe drought during the past two decades. Drought exposure in these countries renders households much more vulnerable to the impacts of political unrest and economic disaster. Drought is not limited to Africa, however. Twelve million Afghanis experienced persistent drought during the first decade of the twenty-first century, clearly exacerbating the conflict in that region. In Asia, 22.6 million people now have inadequate drinking water supplies owing to drought. In India, 130 million have experienced drought during the past two years.

Drought is the single most important weather-related natural disaster. It affects very large areas for months, even years, seriously affecting regional food production, often reducing life expectancies for entire populations, and virtually eliminating the economic viability of countries and whole regions. During 1967 to 1991, droughts affected fifty percent of the 2.8 billion people who suffered from all natural disasters, accounting for the deaths of thirty-five percent of the 3.5 million people who lost their lives to natural disasters. Moreover, building subsidence, engineering projects, and relief measures following droughts involve high costs. During first decade of the twenty-first century—proclaimed the Decade for Natural Disaster Reduction by the United Nations—widespread, intensive droughts were observed on every continent, leading to massive economic losses, destruction of ecological resources, food shortages, and the starvation of millions

To begin a study of this tremendous climate event and how it may be linked to cycles of planetary rhythms, it is useful to sample the years of major droughts in California. Those years are 1911, 1919, 1929, 1947, 1950, 1958,

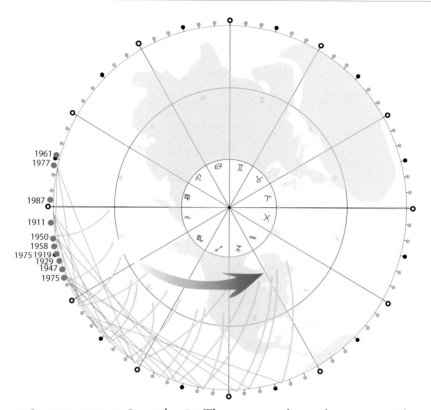

1961, 1975, 1977, 1981, and 1987. These are merely numbers to us until we see them on an eclipse grid chart.

Seen in the context of the eclipse grid, the placement of the eclipse points over the central Pacific for the previous century reveals a remarkably consistent sweep of the jet curves into a berm pattern over the eastern Pacific. In the *berm pattern*, the most likely path of the jet stream is to the northeast (red arrow) as the jet seeks to model the ever-northerly latitude of the jet curves making up the eclipse grid. This track would move storms into the PNW rather than to the south. When the eclipse grid for each position of the major droughts for California is put onto the chart, a thicket of jet curves is formed and oriented to the berm pattern. This striking pattern does not mean that there will be a drought in California every time the eclipse points are over the central Pacific, but the sheer synchronicity and coincidence of the chart shows the potential of the eclipse grid model for studying the severe climate patterns that surround droughts.

DROUGHT CYCLES IN THE DUST BOWL: 1930–1936

The presence of drought conditions in the United States is coupled with profound human suffering. Drought researchers recognize that most prolonged drought conditions in the past were actually caused by poor agricultural practices. Soil moisture evaporates because of the overuse of underground aquifers or improper tillage. The land dries up and retains heat, thus hindering cloud formation. This is frequently the way that drought cycles are perpetuated.

The Dust Bowl, the deepest Midwest drought in the 1930s, was nevertheless a classic meteorological event and primarily the result of unusual climate patterns. Although farming practices aggravated the seriousness of the Dust Bowl, the climatic patterns behind the great drought were devastating in their rhythmic relentlessness.

Four distinct droughts occurred in the Dust Bowl years. The first, 1930 to 1931, was a prelude. The 1934 summer drought deepened the pattern, and the 1936 drought dealt the deathblow to agriculture in the High Plains. The final period, 1939 and 1940, was a lesser event. Because most precipitation in the High Plains falls during thunderstorms in late spring and early summer, the charts depict the eclipse positions in effect in April of the study year. The first chart (next page) shows the planetary situation in spring 1930.

The eastern pair of eclipse points was over western Africa. The western pair of eclipse points was over the mid-Pacific and placed in significant aspects to Saturn. Uranus was close to the two eastern points, with the lunar node between them. Together, the rhythms of Uranus and the node would strongly affect the two eastern points. In the spring of 1930, the rhythms of Uranus and the lunar node supported persistent high pressure on the two eastern eclipse points over the desert.

Together, the four 72° jet curves form a disturbance diamond (yellow) over the High Desert, the High Plains, and parts of the upper Midwest. The western pair of points was under the influence of Saturn at a strong high-pressure value on the solar reflex point. At the time, Saturn was aspecting the lunar point to an intermittent value that results most often to weak or absent manifestations. A pale green line shows the jet curve for this. The strong high-pressure aspects on the 72° jet curves from the eastern pair, and the strong high-pressure value on the 72° jet curve from the solar reflex point put strong

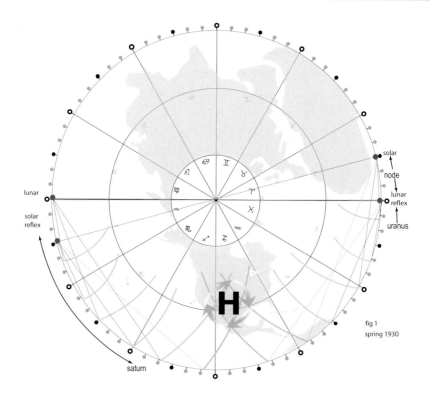

high-pressure values for most of the month of May 1930 over the southern inter-mountain areas and the High Plains. This is the high-pressure area on the chart. The persistent block in this area prevented the low-level monsoon storm jet from moving out of the Gulf of Mexico into the Midwest or High Plains. This was the precise pattern in May 1930. The planetary influences that gave rise to these patterns did not shift much during the months of May and June. There was a kind of counterpoint between Uranus and the node, where any potential low-pressure influence by Uranus moving in arc in late May was offset by the node oscillating on station and reestablishing strong high-pressure values on the eastern points. The node remained on station for the whole months of April and May and half of June. During that time, it repeatedly influenced the eastern points to high pressure.

The space within the eclipse diamond was the exact west-to-east border of the high-pressure ridge that formed in April 1930. The blocking high that brought on the failure of the spring rains covered the area from northern Mexico to Oklahoma, with a strong block reaching northward into the

Dakotas. This pattern coincided with the beginning of the Dust Bowl. By July, the node began moving off station and shifted patterns every ten days. These shifts broke the stagnant ridge and developed more variable weather. However, the two-month dryness had established the beginning of the "creeping drought" phenomenon, in which drought conditions develop over a period of several years rather than in a single year. The next significant drought pattern was in April 1934.

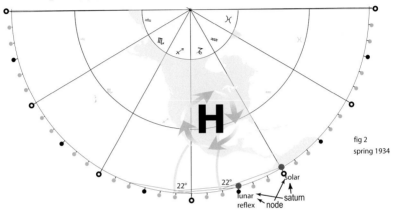

fig 2
spring 1934

The pattern for April 1934 (figure 2) shows strong high-pressure values on the set of 22° eclipse points, which had migrated to the longitude of the Midwest. The chart shows the solar eclipse point in the longitude of Florida and the lunar reflex point in the longitude of Texas. Instead of the 72° jet curves shown operating in the previous chart, this one shows the 22° jet curves from the points. The 22° jet curves are particularly active in the stimulation of northerly surges of tropical air mass movements. We see that the jet curves cover an area from Cabo San Lucas on the Baja California peninsula in Mexico to Oklahoma. High pressure in these areas would most likely support the emergence of low-latitude blocking patterns over the southern High Plains. In both 1930 and 1934, the southwest was the area that first manifested the drought pattern. The dryness spread gradually east as the drought deepened from spring to summer. April 1934 was a time of relentless heat and high pressure. During April, a pattern of twenty-four days developed when either Saturn or the node produced high-pressure values on the 22° jet curve from the lunar reflex point. Saturn or the node produced high pressure on the other 22° jet curve for a total of sixteen days. This means that, for the whole month, there were only a few days when high-pressure values were not over the Southwest. The high-pressure ridge once again fills the Southwest and the High Plains during the spring.

This block extended from the Texas panhandle north to Missouri and Illinois and southeast to the Gulf coast near the Ozarks. This area is the general position of the two jet curves. This pattern continued into July, as the node went on station at a strong high-pressure aspect to the 22° jet curve from the lunar reflex point. Saturn was also at a high-pressure aspect to this point for the whole month. July 1934 was the hottest July on record for Missouri. The clockwise circulation around a ridge placed along the lunar reflex jet curve drew hot air from the Mexican desert into the Midwest while sealing off any Gulf Coast monsoon.

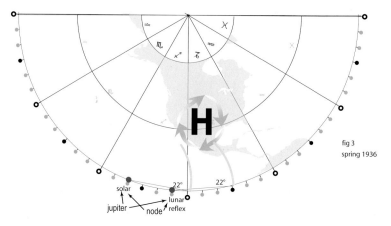

fig 3
spring 1936

The next drought siege for the Midwest and High Plains took place in 1936. Figure 3 shows the eclipse placements for that spring. The points straddled the West Coast. Once again, it was the low-latitude 22° jet curves from the points that formed the site of a low-latitude tropical high-pressure ridge over the southern High Plains. It is clear from the chart that the eclipse points migrated to the west, but the other side of the 22° jet curves from the points is now over the southern High Plains. When these points and tropical jet curves transit the United States, the result is often that the tropical jet stream is enhanced. This was unfortunate for that particular period of six years. First Saturn and then Jupiter worked in concert with the lunar node to create unusual syncopations of high-pressure rhythms on the eclipse points. The set of points moved away from Saturn, only to run into Jupiter, and the whole scenario of nodal interaction played out on the jet curves from the other side of the eclipse points. This was a highly unusual cosmic rhythm.

In the interplay between the node and Jupiter in April 1936, the lunar point fell under the influence of high-pressure values for twenty-six days out of thirty. This means that the jet curve over Texas supported tropical high pressure for the whole month. After two years of searing summer temperatures in this area, the whole climate regime shifted to blowing winds, dryness, and heat. This was the driest summer on record for Missouri, as a ridge established itself once again over the southern High Plains and crept eastward into the Midwest as the summer progressed. The drought continued through the spring and into the summer, with Jupiter and the lunar node playing high-pressure tag team with the two eclipse points. In late July, two eclipses gradually phased out the Jupiter influence as the points moved west into a position near Hawaii and out of the tight dance between Jupiter and the node that characterized the spring and early summer. After the eclipses in July, some ameliorating rains began to fall on the parched areas, but the destruction and social displacement caused by the relentless dryness had taken their toll on the soul of a nation.

It is interesting to note that this section of the country also experienced a crippling drought during the period from 1950 to 1956. It may be more than coincidental that, during those very years, the eclipse points traversed exactly the same regions that were so instrumental in the formation of the Dust Bowl. It is also interesting to note the presence of Saturn, Jupiter, and the lunar node adjacent to the eclipse points in the formation of those patterns of high pressure across the southern High Plains.

AMERICAN HIGH PLAINS DROUGHT PATTERNS: 1988 AND 2007

To climatologists, the period in North America from April to late July 1988 is "The Great Dry." Beginning in spring and continuing into late summer, it involved the severe placement of a set of blocks, or ridges, that prevented the onset of moisture currents from the Gulf of Mexico, which are the source of much of the summer thunderstorm activity in the Corn Belt and the wheat-growing regions of the United States.

There are many causes of drought in the United States Midwest and High Plains. Much research has gone into understanding the climate patterns that periodically parch that growing sections of the country. In general, the most fundamental climate pattern is the presence of a blocking high-pressure area over the Western U.S. during the spring and summer months.

Climatologists recognize that a nineteen-year rhythm is often linked to the presence of drought and crop failure in the High Plains. Midwest drought cycles are not closely linked to this. However, even in the Midwest, Central States, and Corn Belt, droughts usually originate with some form of blocking pattern over the High Plains. Such droughts then spread eastward during the season as the blocking ridge builds with each day that the rains fail to cool the earth.

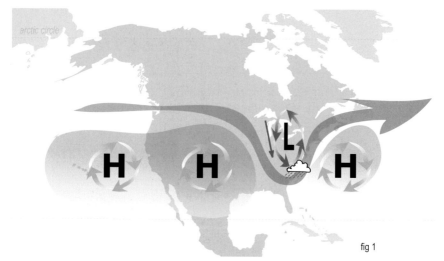

fig 1

Figure 1 shows this extensive High Plains blocking pattern in the context of the three major high-pressure areas that contributed to the Great Dry of

1988. In the chart, a large dome of high pressure sits over the mountain states and the High Plains. The circulation around this area is clockwise, which means that, on the eastern leading edge of the dome, air is moving from north to south. The small black arrow depicts this motion. The main cause of rain in the Midwest during the summer is moisture moving from south to north out of the Gulf of Mexico. During rainy years, this moisture forms a low-level jet of moist air flowing northward from Brownsville, Texas, which acts as the source for thunderstorms that bring life to the soybeans, corn, and other crops grown in the Corn Belt in the Great Lakes area. However, when a strong high-pressure area establishes itself over the High Plains, the north-to-south circulation diminishes the monsoon flow from the Gulf of Mexico, bringing drought. Frequently, in the most serious drought conditions, there are two more high-pressure areas near the United States. One is over the eastern Pacific, the other is over the western Atlantic Ocean. These can be seen in figure 1. The placement of these highs creates a narrow trough over the East Coast, where wet conditions prevail. We will say more about these other highs as we go along.

In figure 2 (following page), an expanded chart shows a more comprehensive view of a *tele-connection*. We see that the three highs implicated in serious drought are actually part of a whole series of highs and lows connected over vast distances. These super-remote linkages of air masses are teleconnected to each other, similar to waves in the ocean. In the chart, we see that the highs over the United States originate near Asia. The green snake-like arrow is the jet stream winding its way around the highs, forming lows, or troughs, in the breaks between the high-pressure masses. Climatologists designate this pattern of highs and lows stretching across the Pacific the "North Pacific Index" (NPI) pattern. The rhythm of the flow across the Pacific in the NPI pattern forms a consistent sequence of lows and highs that follow one another, except in the area between Hawaii and the Aleutian Islands. Geometrically, by following the sequences of highs and lows, we would expect a trough or low to form there if the spacing of the waves of air masses were consistent. Indeed, such a trough would normally form between Hawaii and the mainland, and the result would be rain for the West Coast, since a low would form off the coast and guide storms into the coast. If this pattern were unfolding in winter, the high over the eastern Pacific would normally be much weaker, and the gentle saddle shape between the high in the mid-Pacific and

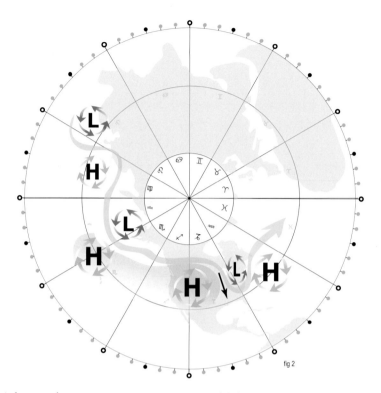

fig 2

the high over the inter-Mountain areas would dissolve and allow a low to dig south along the West Coast.

However, this is a chart for late winter, and the pattern is forming for the emergence of high pressure in the eastern Pacific. During early spring, this pattern has the Hawaii high beginning to expand and move northward, blocking the jet stream from dropping to the south. This is the first element in the formation of this drought in 1988. The summer regime is supportive of a stronger eastern-Pacific high, which prevents a trough from forming off the coast. However, there is not necessarily a major drought each summer when the Hawaii high strengthens, meaning there must be more to it than this element alone, and indeed there is. However, the position of the Hawaii high in the spring is significant in this particular case, as we shall see. We could say that the growth of the Hawaii high during the summer is the fundamental ground of the drought sequence of 1988 in the High Plains.

The next image (above) shows the same chart for the North Pacific Index (NPI) pattern across the Pacific, but superimposed on a chart for the eclipse

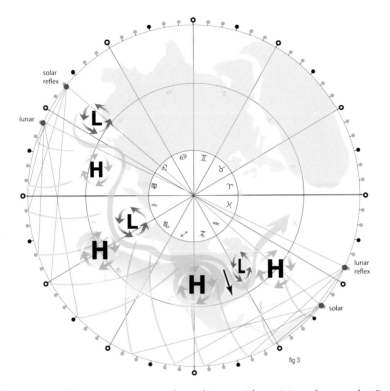

fig 3

grid patterns. Here, we can see the eclipse grid positioned over the Pacific. During this year, the eclipse grid stretches from the United States East Coast all the way out to the East China Sea. It is interesting to note that the tele-connection pattern between the inter-mountain area of the western U.S. and Indochina is an image of a berm pattern in the eclipse-grid model. The geometry of the eclipse grid is harmonic to the trans-Pacific tele-connection between Indochina and the United States East Coast.

The low over Southeast Asia is in the bottom of the berm. The high in the longitude of the Sea of Okhotsk, east of Russia, is over the 22° jet curves. The trough over the Bering Sea centers on the two 45° jet curves. The two ridges that link from Hawaii to Denver are on either side of the highest part of the berm, the disturbance diamond (yellow).

In the next chart, figure 4, thicker lines accentuate the 72° jet-curve lines that form the disturbance diamond (yellow). The patterns of high pressure around this diamond support the maintenance of high pressure in the saddle between the tele-connected highs in the eastern Pacific and the inter-Mountain

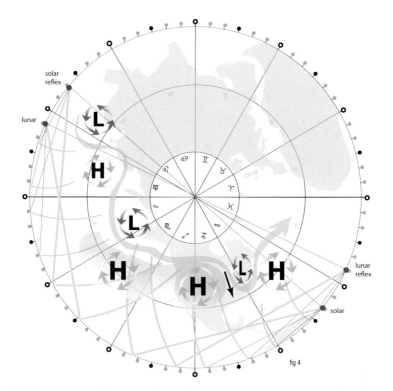

fig 4

states in the summer of 1988. When these 72° jet curves are stimulated to high pressure, the most predictable result is that high pressure dominates high-latitude areas near the top of the 72° jet curves. Here, this is the area to the east of the 72° jet curves from the western pair of eclipse points (western Canada) and the area to the west of the eastern pair of 72° jet curves (Bering Sea). The eclipse diamond is active in this chart, maintaining high pressure across the saddle between the mid-Pacific and the inter-Mountain areas.

In figure 5, the 45° jet curves from the eclipse grid are accentuated. We see that the eastern pair supports high pressure over the inter-mountain areas, while the western pair supports high pressure over the mid-Pacific. Together, the eclipse lines, the 45° jet curves, and the 72° jet curves would be in perfect position to support high-pressure dominance from the China Sea to the Bermuda high if the eclipse points in the east and west were aspected to high pressure by planetary motion. It turns out, however, that in 1988 the planetary motion was very much in support of sustained high-pressure values on both sets of eclipse points for long periods during the spring and early summer

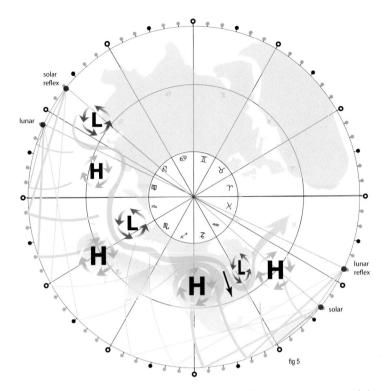

fig 5

of that year. This means that high pressure would dominate most of the areas depicted in the preceding charts, supporting the high-pressure phase of the North Pacific Index. The reason for this overwhelming high-pressure influence can be seen in the next chart.

In figure 6, the position of Mars and the lunar node are in the eastern pair of eclipse points. In this sequence, the flow between Mars and the lunar node was such that the motions of the planets between the eclipse points stimulated high-pressure values at almost every motion-in-arc event. This means that, when Mars moved to a high-pressure aspect between both points, high pressure values were generated on both of the eastern points. In 1988, the mathematical relationship between the two eclipse points supported the formation of high pressure on each point for extended periods. This rhythmic phenomenon kept high-pressure values on the two eclipse points for most of the early summer of 1988. In the eclipse grid model, this was a strong signal for drought on the High Plains at that time. It is a highly unusual cadence for the movements of the planets, and the unusual nature of the extended drought

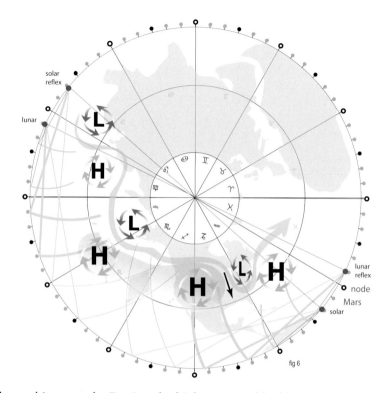

solar
reflex

lunar

L

H

L

H

H

L

H

lunar
reflex
node
Mars
solar

fig 6

links to this anomaly. During the high-pressure blocking event in 1988, two planets, Mars and the lunar node, were stationed between the two eclipse points for an extended period during the summer.

Figure 7 shows the chart for summer 2007. Between 1988 and 2007, the lunar node transited the entire zodiac to approximately in the position it was in during the summer of 1988. In the eclipse grid model, this return positioning constitutes an analog year and is considered the source of the nineteen-year-plus rhythm of the High Plains drought cycle. This means that the return of the nodal positions is the same as it had been approximately two decades earlier. In the eclipse grid model, this nodal return forms the rhythmic basis for the choosing of an analog pattern for a particular year. This method attempts to explain the well-known climatic phenomenon of decadal or inter-decadal influences, which is the return of particular climatic phenomenon in a ten- or twenty-year cycle. In Doc Weather (www. docweather.com), these decadal cycles relate to the retrograde motion of the lunar nodes. Similar to the chart in 1988, the chart for 2007 reveals the

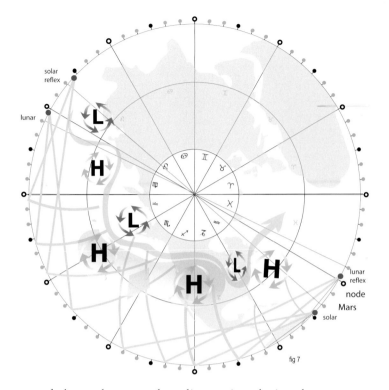

fig 7

placement of planets between the eclipse points during the summer months. This should prove to be a consistent feature in the establishment of long-lived high-pressure areas that evolve into an NPI pattern across the Pacific. This would most likely end in a blocking high over the High Plains for the early summer. Mars and the lunar node and Uranus will be found within the two eclipse points, promising a prolonged blocking pattern for the Midwest monsoon. The dryness should be in place early, as Mars transits the points in the early spring. Later, in July, there is a strong series of high-pressure relationships between Uranus and the lunar node that should bring memories of 1988 into the minds of older growers in the High Plains and the Midwest.*

* Note: this section was written before the event. For a study of summer 2007, please see chapter 10.

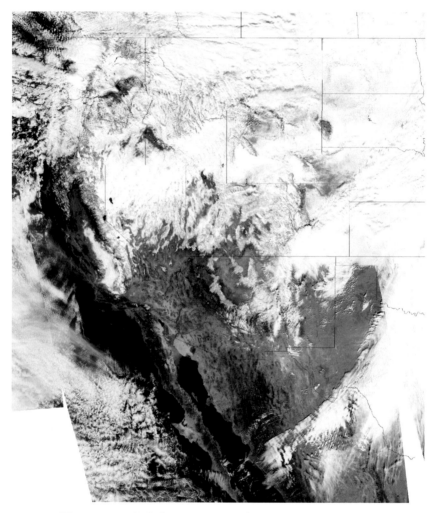

Winter storms lash the western United States, January 12, 2005
(NASA Satellite Image)

CHAPTER 8

FLOODS

RECORD RAINS IN SOUTHERN CALIFORNIA: WINTER 2005

Record rains in Southern California during the winter of 2005 were part of a blocking pattern, an *omega block*. The term comes from the Greek letter *omega*, which looks like a horseshoe, open at the bottom (Ω). An omega block forms when a high-pressure area in a low-latitude position surges to the north, forming a long vertical ridge. At the same time, two low-pressure areas dig into the south on either side of the surging ridge. The lows lock in place to the east and west of the ridge. The jet stream flows south to the first low, then to the north again around the high-pressure area, and then south again around the second low. The patterns around this circulation tend to stabilize as these features feed off each other. The average omega block lasts five to seven days.

Figure 1 (next page) shows an omega block that formed over the eastern Pacific in January 1998. Its features result in a pattern in the eastern Pacific, in which the western low draws air to the south from the Gulf of Alaska and then feeds it rapidly to the north, usually with an abundance of water. This occurs because the cold polar air tends to pick up moisture over the ocean, which is then shunted north, where it cools and forms clouds and fronts. Then, the jet stream once again digs south on the eastern leg of the omega and usually deposits the gathered moisture into the dangerous semicircle of the eastern low, which is trapped against the coast. The ridge between is usually the site of very dry weather, while the two lows are often the sites of abnormal precipitation. After a week or so, the block usually dissipates and drifts to the east.

From December 30, 1997, to January 5, 1998, an omega block in the Gulf of Alaska created strong rains for Northern California and the PNW. The eclipse positions for that event projected the disturbance diamond into

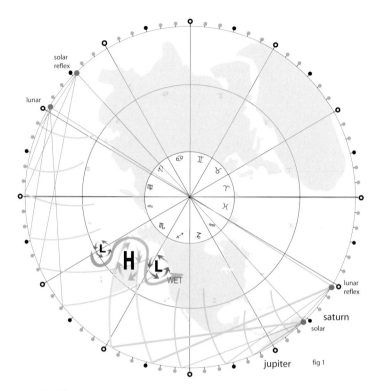

fig 1

the central Gulf of Alaska. Both Saturn and Jupiter were influencing the 72° jet curves from the eastern pair. On December 30, 1997, Jupiter moved in arc, putting a high-pressure influence on both 72° jet curves. On that day, a ridge grew from the longitude of Hawaii north and west toward the Aleutian Islands, while a low dug to the south in the longitude of the two 72° jet curves. On the next day, December 31, Saturn moved to high pressure, also on both of the eastern eclipse points. This surged the ridge again and stabilized the low that had dug south. For the next four days, the ridge from Hawaii to the Aleutians was the site of strong high pressure as a low formed over the central Pacific to the west of the ridge, and a low formed over the eastern Pacific to the east of the ridge. The western low brought heavy rains to the PNW and the Sierras. On January 4, Jupiter moved in arc and brought a low-pressure aspect to the 72° jet curves. The ridge faltered on January 5, and then, on January 6, it pinched off in the north and drifted into the polar regions. What remained of it collapsed farther to the south and drifted east as the pattern broke down. During this event, the ridge occupied a space that was coincident

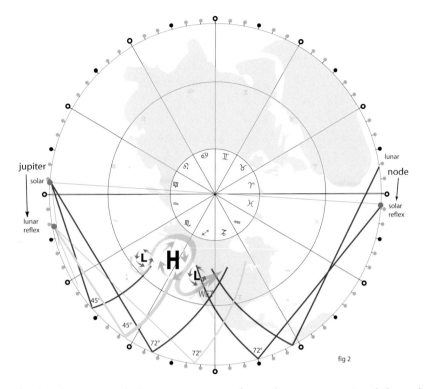

fig 2

with the placement of the 72° jet curves from the eastern pair while under high-pressure aspects. This example introduces the idea of an omega forming under the influence of the eclipse grid.

Figure 2 shows the eclipse grid pattern for the 2005 flooding in Southern California. We see that the important nexus of the four 72° jet curves is placed over the mountain states, where the 72° jet curves tend to have a strong impact on continental weather and a reduced impact on coastal weather. However, the 45° jet curves over the eastern Pacific are closer to the coast during 2005 than they were in 1998. This means that any blocking pattern on the western pair of eclipse points over the mid-Pacific will produce a block into the central Gulf of Alaska rather than the western Gulf of Alaska. An omega pattern in the central Gulf of Alaska would bring a strong jet due south with a low trapped on the eastern leg securely against the West Coast. That was the dominant pattern of the unprecedented rains in Southern California during 2005. Similar to 1998, in 2005 Jupiter was again a factor in forming the flooding and blocking patterns. In contrast to

1998, however, this time it was placed over the western Pacific, from where it influenced the western pair of eclipse points. The two 45° jet curves over the eastern Pacific are part of that influence.

Figure 2 also shows the planetary influences from Jupiter in the west and the lunar node in the east on January 7, 2005. From this position, Jupiter established a strong high-latitude, high-pressure area over Alaska as it aspected the lunar reflex point to high-pressure values. This influence manifested as a block on the 45° jet curve from the lunar reflex point (thick blue). This aspect led to a high parked over the eastern 45° jet curve. At the same time, Jupiter aspected the solar reflex point to low pressure. This aspect manifested on the 45° jet curve to the west as a strong tendency to low pressure (red 45° jet curve). A low formed on this curve. The interaction of the two 45° jet curves set the stage for an omega block. At the same time, the 72° jet curve from the solar reflex point was also aspected to low pressure (red curve running southwest to northeast across Southern California). This placement was coincident with the jet stream in the eastern Pacific dropping far to the south.

From the eastern pair of eclipse points, the node had aspected the 72° solar reflex jet curve over the western states to low pressure (red 72° jet curve over the mountain states). This curve ran generally parallel to the West Coast, from Baja to Vancouver. A low formed on this curve off the coast in the longitude of Vancouver, where the curve crossed the mountains out into the ocean. The low there received the jet from the high over Anchorage, and the circulation of the low directed very cold fronts down the coast and into Southern California. This block with the accompanying locked low-pressure areas formed on January 7 and held on until January 11. Meanwhile, violent storms brought stunning levels of snow and rain to the south state. In Reno, this was the biggest system to hit that city in eighty-nine years. Eventually, the high moved off its position because the Moon was transiting the West Coast on the January 11. Jupiter, aspecting the 45° jet curve from the lunar reflex point, kept the block intact from a position near the western points, while the node near the eastern points kept low pressure dominant over the West Coast. Taken together, the two planets provided a perfect patterning for an omega block.

In addition, the rapid shifting of Saturn, Mars, and the lunar node dominated the next two weeks by forming a series of changing aspects that supported moving ridge patterns rather than blocking highs. Following this pattern on January 28, Jupiter moved in arc to a new degree that also supported moving ridges on the two 45° jet curves over Alaska. As a result, any highs that emerged in the eastern Pacific from January 12 to 28 would surge briefly to the north, block for a day or two, and then move eastward instead of locking into position. This shows that there are particular combinations of aspects that must come together in time to produce an omega block. When the components are this strong and persistent, weather patterns are the result.

FEATHER RIVER NEW YEAR'S FLOOD STORM SEQUENCES
DECEMBER 1996 AND JANUARY 1997

The sequences of events leading to the Feather River floods in California in late December 1996 and early January 1997 offer a classic case study of the influences of planetary motion on the "storm gate" in the Pacific Northwest (PNW). That year, a major rain pattern set into the PNW, and floods in Northern California at the New Year of 1997 were part of the unusual rain patterns for the Northwest in the fall of 1996.

The sequences for the Feather River New Year's floods began December 13, 1996, with movements by the lunar node aspecting the western eclipse points. These aspects put a strong high-pressure value on the 72° jet curve from the solar point in the western Pacific (figure 1, blue curve). The 72° jet curve was right against the southern coast of California and extended into the Rockies. This was a significant placement for a strong high-pressure impulse. On the next day, December 14, Saturn moved to low-pressure values on both of the eastern pair of eclipse points (red lines). This put strong low-pressure values on both of the 72° jet curves from the eastern pair of points (red curves just off the West Coast). The resulting very broad low-pressure area dominated the eastern Pacific, up into the eastern Gulf of Alaska. With the disturbance diamond straddling the coast, the patterns for high and low pressures become organized according to standard climatology. In this case, the three 72° jet curves over the ocean were designated low pressure, and the lone high-pressure 72° jet curve was over the Rockies. That put the broad low to the west over the ocean and the block over the coast.

A low that had remained stationary in the western Gulf of Alaska now began to drift to the south as its circulation came under the influence of the predominantly low-pressure areas of the disturbance diamond in the eastern Pacific. The lone high-pressure value of 72° jet curve from the solar point in the west was a very strong value since both Pluto, which was also near to the eclipse points, and the node moved to high-pressure aspects to the solar point on the same day. This type of coincidental motion-in-arc event is most often at the source of unusual weather sequences. When Pluto and the node moved, an intense but small high formed in the disturbance diamond near the 72° jet curve ("H" in blue area) and wedged against the Sierras to the

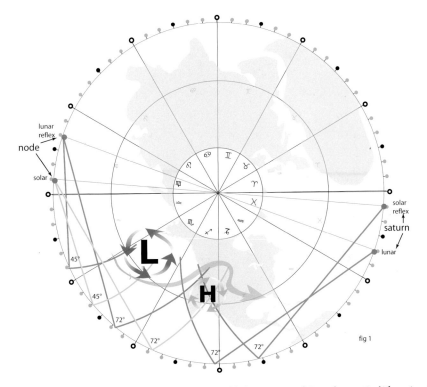

fig 1

south. The center of the high was just off the coast of Southern California. It was intense but was wedged between three other 72° jet curves aspected to low pressure. This meant that it was unable to surge and expand but could block, and block it did.

A low that had formed on the 45° jet curve from the lunar point in the western Gulf of Alaska dropped to the south from its position in the Aleutians. The block against the coast stopped the eastward motion of the low, and it remained over the ocean for the next few days as it gathered strength. Sagging toward the coast during the next few days, the high against the coast was compressed into a narrow ridge running north into the PNW along the 72° jet curve from the western solar point.

On December 17, Jupiter moved in arc shifting the eastern pair of eclipse points to a split between high and low pressure. Jupiter was very remote from the eastern pair at this time. However, it was the only planet moving within 90° of these points. This made it active in the field of the

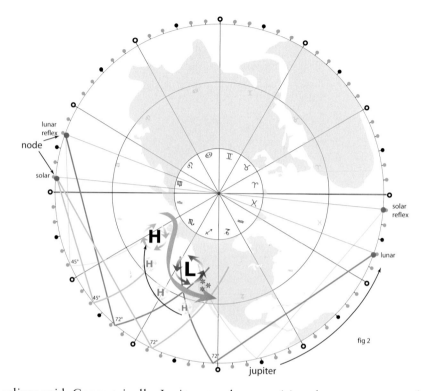

fig 2

eclipse grid. Geometrically, Jupiter stood at a position that was resonant for creating the split harmonics on the grid. The shift was weak, but it changed the pattern in the disturbance diamond in a significant way. Specifically, the 72° jet curve from the solar reflex point changed to a weak, or intermittent, geometric value (pale green thin line). Most often, when this happens a situated block moves off an established position. The shift dislodged the block against the coast and drifted to the west on the easterlies because of its position at low latitude (black curved arrow). Usually, a high that is dislodged in this manner will seek the nearest high-pressure jet curve as it tries to make its way toward the pole. All high-pressure areas in the northern hemisphere move toward the north pole from a low latitude, where they originate, and all lows in the northern hemisphere move toward the equator as part of the climatology of the northern hemisphere. When the geometric numerical value shifts on a given day, established highs tend to migrate to the strongest high-pressure jet curve in their vicinity. This has been observed many times.

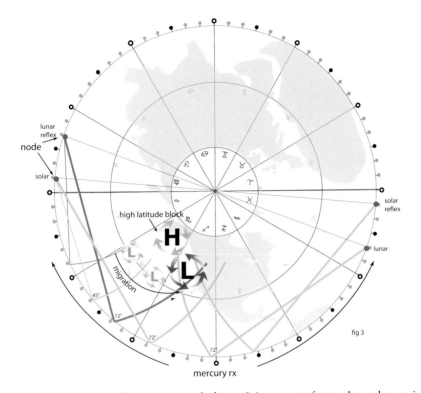

fig 3

As the high moved west toward the 45° jet curve from the solar point, that area became the source of a surge of high pressure going up into Alaska. When the dislodged high moved northwest toward the 45° jet curve from the solar point, the deepening low that had been building was no longer blocked from moving east. The large low swung eastward from the western Gulf of Alaska, and the circulation between the high and the low pulled cold air down from the north (large blue/red arrow). As the storm front made its way up the slopes of the Sierras, the cold polar air converted the warm moisture in the low into a snowstorm. This brought a curious combination of thunderstorms that produced abundant wet snow. This was an image of the hybrid nature of this system. The lightning flashed and the wet snow fell and blanketed the mountains on December 22 and 23.

As the snow fell in the eastern Pacific, a profound shift was happening in the mid-Pacific. The high that had migrated northwest to the 45° jet curve in the middle of the Gulf of Alaska found support for a large surge. This support came in the unusual placement of a Mercury station on December 22 that, geometrically, put strong high-pressure values on both the eastern solar reflex point and the western solar point. Mercury on station split the points. With this shift Mercury stood on station at an angle of 90° to both of these points. It is rare for a planet to go on station at such a sensitive place. A planet on station at a geometric value of 90° to both of the points will generate a very strong northern high-pressure surge on the jet curves under its influence. It also happened that this shift put a high-pressure value on the other eastern eclipse point, the lunar reflex point. This meant a sudden enhancement of high pressure on both of the 72° jet curves from these points. These high-pressure jet curves were in the eastern Gulf of Alaska. A strong growth of high pressure from the western states pushed northward along the coast range, surged northwest into Alaska, and then joined the ridge over the Aleutians, making it a high-latitude-blocking ridge over Alaska.

A few days earlier, the node over the western Pacific had shifted to a strong low-pressure aspect on the lunar reflex point (red lines). This put a strong low-pressure value on both the 45° and the 72° jet curves from that point. At the time, a strong low that had been forming in the Bering Sea now formed a front that sagged south from the low, migrating toward the low-pressure value on the 72° jet curve from the same point (thin black arrow). The surging high pushing up the coast enhanced the high-latitude block in the Gulf of Alaska and forced the migrating low to the south, where it picked up abundant moisture over the ocean. For the next few days, the front approached the West Coast but could not pass easily into the coast because of the high-pressure values on the two 72° jet curves still caught in the surge of high pressure to the north. The front slowed as it approached the coast, and waves of storms spun into the coast bringing abundant moisture with them.

The low became what climatologists call a "Pineapple Express," directing warm fronts filled with moisture from Hawaii onto the West Coast. As the storm came ashore, the warm rains falling on the wet, newly deposited snow from the previous storm created a precarious condition. The fresh snow quickly melted in the warmth of the new storm. Torrents of rain cascaded from the slopes of the precipitous Feather River Canyon, as a wall of water formed in the canyon and broke on the town of Chico in the Central Valley, bursting dams and levees and inundating farmland for miles. The second storm hit the coast in full force on New Year's day and created heartbreak for many families over the holiday season.

Had those storm cycles been reversed, a wet Pineapple Express storm would have filled the reservoirs, while the colder, snowy storm would have settled into the mountains, providing a good water reserve for the following summer. As it turned out, most of the water ran down into the valley and into the ocean without being adequately preserved. The syncopated motions of the planets had created a situation in which the atmospheric responses to motion-in-arc events along the eclipse grid formed a rhythmical sequence for the history books.

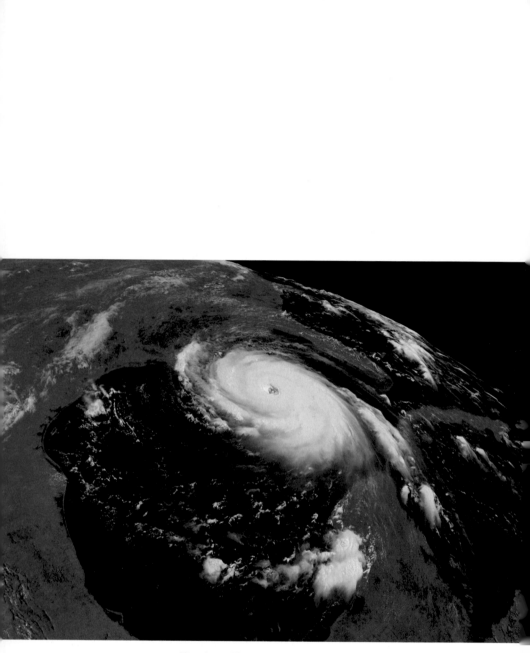

Hurricane Katrina, August 28, 2005
(NOAA Satellite and Information Service)

Chapter 9

HURRICANES

Hurricane Analysis with Planetary Motion

Using the techniques of the eclipse grid model, it is possible to get an interesting take on the likelihood of hurricane frequency and general geography. The first chart (following page) shows the positions of major hurricanes in 1998. Classic Cape Verde storms such as Hurricane Danielle start near the Cape Verde Islands off Africa as a wave of air coming from the African continent about every ten days during autumn. The wave then travels the tropical easterly winds (red arrow) toward the U.S. If conditions are good for propagation, the wave is enhanced and a hurricane evolves. If conditions are not supportive the wave fizzles out and no storm results.

What supports a wave to grow into a hurricane? This is the central question. When the tropical easterlies actively blow from east to west across the tropical Atlantic, hurricanes can expand. When the easterlies are weak or when a westerly wind blows in the tropics, the waves moving off Africa are suppressed, and the storms themselves are sheared off as they are in the process of building. Of course, these are just two of the many factors in hurricane development and tracking, much of which remains a mystery to science.

In line with the eclipse grid model, this short study used the direct and retrograde motion of planets over the continental United States and the Atlantic as a backdrop for plotting major storms between 1998 and 2007. The study shows a remarkably suggestive coincidence between the positions and transits of planets in particular longitudes, as well as the frequency, intensity, and onset of major storms in the Gulf of Mexico and the western Atlantic.

In 1998, Jupiter and Saturn were retrograde (east to west) near the Cape Verde Islands. These movements were coincident with two classic Cape Verde storms between July 16 and August 16 that year. The chart shows the flow of the easterlies from the eastern Atlantic as a precursor of the motion of

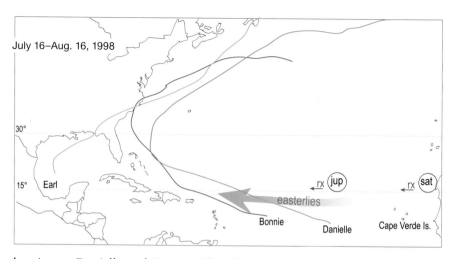

hurricanes Danielle and Bonnie. The idea is that the flow of the retrograde motion of Jupiter and Saturn in the locale of the Cape Verde Islands is coincident with a stronger-than-normal flow of tropical waves from that area off the west coast of Africa.

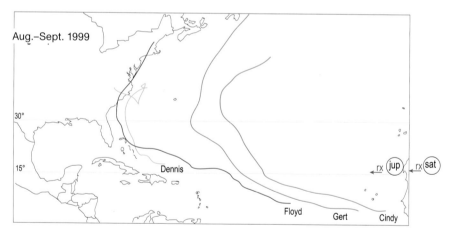

In the chart for 1999, Jupiter and Saturn are still in the eastern Atlantic during their retrograde periods, but both of them are much farther to the east than in 1998. Three Cape Verde storms evolved, but two of the tracks (Cindy and Gert) are shifted to the east of the storms in 1998. This is coincident with the eastward shift of Jupiter and Saturn. Floyd is also considered a Cape Verde storm, but its initial starting position allowed it to make a U.S. landfall. The influence of the two retrograde planets off Africa during August and

September of 1999 was once again coincident with the generation of classic Cape Verde storms.

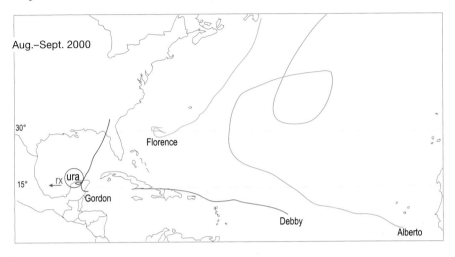

Aug.–Sept. 2000

30°

Florence

15°

rx ura

Gordon

Debby

Alberto

By summer 2000, both Jupiter and Saturn were over eastern Africa. This position exerted much less influence over the eastern Atlantic easterly winds in the Cape Verde area. With little drive to the west, Alberto, a Cape Verde storm, has a confused track, as does Florence. Of note is the presence of a retrograde loop of Uranus over the Yucatán Peninsula, the site of the looping track of Gordon. The idea is that, lacking the thrust of Jupiter and Saturn in retrograde motion off West Africa, the tracks of the storms loop and meander.

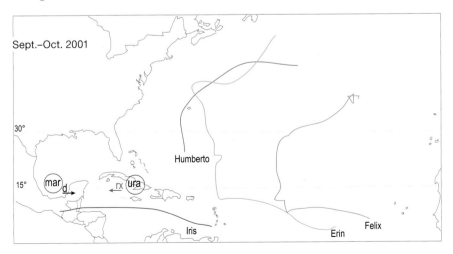

Sept.–Oct. 2001

30°

Humberto

15°

mar d

rx ura

Iris

Erin

Felix

In 2001, major hurricanes were confined to the middle of the Atlantic, with a hesitating gesture to hurricanes Erin and Felix. No strong easterly push is present in the eastern Atlantic to drive storms in a classic recurve shape over the Gulf Stream in the western Atlantic. Iris, in the Caribbean Sea, was coincident in time with the retrograde movement of Uranus. Mars entered the Gulf of Mexico in October. A Mars transit in which the planet moves rapidly from west to east is often coincident with a strongly diminished easterly component to the winds that build hurricanes.

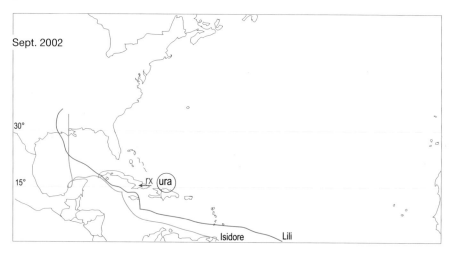

The chart for 2002 is very instructive. The only major storms in that year were in the longitude of the retrograde Uranus over the Dominican Republic. Mars, in a two-year orbit, was not a player in 2002. It is notable that Neptune was about to enter the western Gulf of Mexico late in 2002. That would put two planets moving retrograde in the Gulf of Mexico in the next year.

Active west-to-east Mars movements through the Gulf of Mexico in 2003 coincided with diminished easterly winds in August and September 2003. With diminished easterly winds, the storms sheared off before they could build strength. As Mars transited the eastern Gulf of Mexico in July, it slowed to a stop mid-month over Cuba. At that time, Hurricane Claudette formed in the Gulf of Mexico. By the end of July, Mars was moving retrograde over Cuba. This was coincident with the generation of Hurricane Erika, which moved from the Caribbean Sea into the Gulf of Mexico and tracked to the west of the retrograde Mars. In that season only one major hurricane, Fabian,

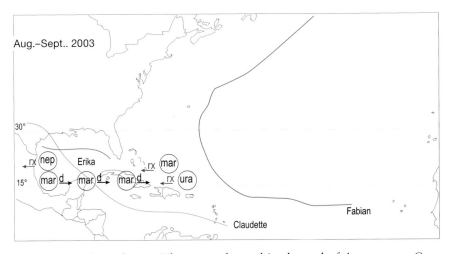

Aug.–Sept.. 2003

developed in the Atlantic. The storm formed in the end of August as a Cape Verde storm. The generation of that hurricane was coincident with the strongest retrograde period of Mars. On September 4, Fabian veered north in the longitude of Mars, over the Dominican Republic, just as Mars was slowing to a stop to shift back to direct motion.

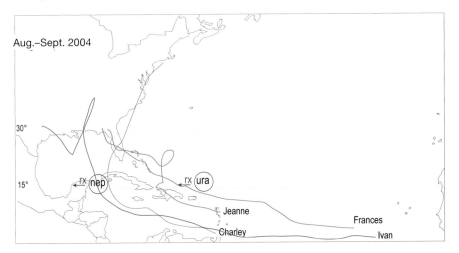

Aug.–Sept. 2004

The 2004 hurricane season was memorable. Four major storms made landfall in the continental U.S. We can see in the chart that the retrograde motions of Uranus and Neptune precisely bracketed the portal they used during the height of the storm season. Because Mars was in the Asian side of its two-year orbital period, no west-to-east Mars influence interrupted the buildup of easterly winds across the Atlantic. This was coincident with an enhanced easterly flow that brought many storms across Florida, exactly between the two retrograde outer planets at the height of their east to west motion.

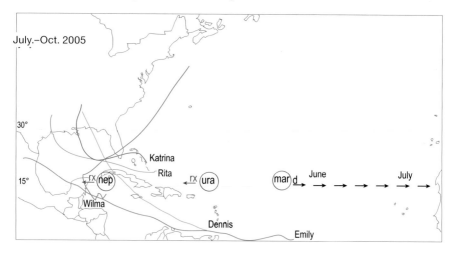

Once again, in 2005, precisely the retrograde motion of Uranus and Neptune bracketed a remarkably intense hurricane season. To be sure, other hurricanes formed outside the brackets, but the real action was between Neptune and Uranus. Mars was in the Atlantic side of its two-year orbit, but it transited the Gulf of Mexico in April and the western Atlantic in May. As a result, it was over the Cape Verde islands in July and out of the picture. It takes about a month for slow-moving planets to begin actual progress in the retrograde period. Thus, it was not until mid-July that Uranus and Neptune were both well established in retrograde motion. At that time, the western Atlantic between the two planets became a hotbed of hurricane activity. During that time, too, the area of the western Atlantic and the Gulf of Mexico was the site of the remarkable buildup that gave rise to Hurricane Katrina.

Because 2004 and 2005 were back-to-back extreme hurricane seasons, most forecasters predicted a repeat of that intensity in 2006. Nevertheless, that hurricane season was relatively calm. Two major storms turned north in the mid-Atlantic. What happened? Were Neptune and Uranus no longer over the western Atlantic? They were, but another element had entered the picture from the east: the ascending lunar node. In the years between 1998 and 2007, the lunar node was not present in the Atlantic. It was over Africa

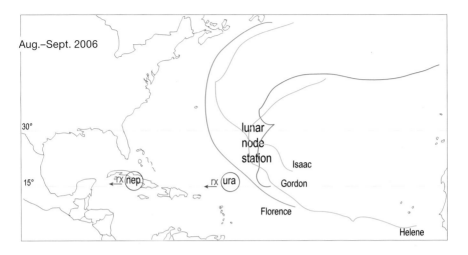

in 2005, but it was not on station at that time. The rhythm of the node is such that it constantly alternates retrograde and direct for most of the year. Only at the time of an eclipse does it stand still for a month-and-a-half. In August and September, the lunar node was on station because of the two eclipses in September. This in itself is not significant for hurricanes, because it happens every six months. However, in 2006 the station of the lunar node took place in the middle of the Atlantic Ocean during the height of hurricane season, August and September. The station was in the longitude of the turning points of hurricanes Gordon, Helene, Florence, and Isaac.

These phenomenological descriptions are not offered as cause-and-effect proof of hurricane genesis and tracking, but they do point to remarkable coincidences that will, one hopes, serve as food for thought about the relationship between planetary motion-in-arc and climate.

The Stillman Creek flooding a cornfield in north-central Illinois, 2007

CHAPTER 10

CLIMATE SYMMETRY EXPERIMENT, 2007

The study outlined here is the result of a summer-long phenomenological experiment undertaken to research a possible symmetrical relationship between weather systems in the United States and those in the United Kingdom during the summer of 2007. The experiment ran from spring through to early autumn of 2007, because a particular symmetrical geometric interrelationship among Saturn, Jupiter, and the lunar node manifested in that period. As expected, owing to unusual angular aspects between Jupiter and the lunar node in relation to the eclipse points, a drought pattern was going to set up that summer for the U.S. Midwest. For about seven weeks in July and August 2007, the influences of Jupiter and the lunar node would be in synchronous, harmonic angular aspects to the eclipse points over the western Atlantic. A geometry composed of planetary angular aspects related to eclipse points determined the harmonies. The geometric structure used for this work is the "eclipse grid."[*]

The last time a similar synchronous geometric event happened of this import was in 1988, when a deep drought developed over the U.S. Corn Belt. In 1988, eclipse grid patterns similar to those for 2007 arose between the eclipse points, which are the fundamental source of the eclipse grid. Studies of the weather patterns in the 1988 drought pointed to the potential for a recurring climate pattern in 2007. In those studies, the activity and placement of Jupiter showed a strong correlation to the timing and placement of a continental blocking ridge that caused the 1988 Midwest drought. The only seasonal difference between 1988 and 2007 was that, in 2007, the expectation was that the drought would begin in July, whereas the 1988 drought pattern began in May. Other differences in the geometrical influences acting on the

---

[*]   An explanation of the source of those harmonics on the eclipse grid is beyond the scope of this chapter. For background on these methods and the eclipse grid, please refer to the numerous articles on the Internet at docweather.com and climatrends.com.

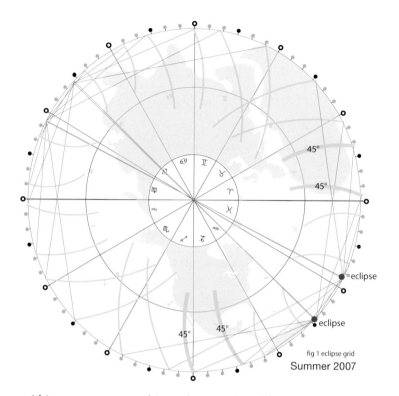

fig 1 eclipse grid
Summer 2007

eclipse grid in summer 2007 ultimately proved problematic. Unusual weather patterns during that summer in both the U.K. and the High Plains area of the U.S. were evidence of those differences. Those unusual weather events are the focus of this chapter.

Working with the eclipse grid also created a unique opportunity to observe potential symmetries in weather patterns in the U.S. and the U.K. for summer 2007. In the process of setting up the symmetry experiment, the collaboration of Jonathan Code of Gloucestershire, Stroud, U.K., was recruited to provide a close watch on the weather patterns in England during the expected drought sequences for the U.S. Midwest. Figure 1 illustrates the basic eclipse grid, the basis of the symmetry experiment. The initial idea was that the summer 2007 eclipse grid, with its points over the western Atlantic, would place harmonic sets of 45° jet curves over both the western third of the U.S. and an area just off the coast of the U.K. A jet curve is a zone in which the grid geometry often shows enhanced potentials for air mass development and placement. The jet curves (shown in green curves) are a very useful tool for making long-range forecasts. This placement of the eclipse points and jet curves created the potential for a geometrical symmetry between the High Plains and the area near Iceland off the British coast during the summer of 2007. It was thought that the symmetry of this particular analysis of the eclipse grid would provide a way to assess the influences of the lunar node and Jupiter on blocking patterns at a hemispheric level.

With this particular grid arrangement, planets influencing the eclipse points would simultaneously influence both sets of 45° jet curves (thick green curves). This would allow a comparison to determine the existence of symmetry in the timing of events. A further impetus for this study was the fact that documenting the patterns would be easier, since both Jupiter and the lunar node would be on station (not moving in longitude) at high-pressure values to these eclipse points from early July through late August. Reason assumed that high-pressure ridges on both sets of 45° jet curves would dominate this period of time. By watching for fluctuations on both sets of curves as they responded to the brief and intermittent motions of other planets, which also moved against these points, one assumed that a symmetrical relationship in air masses could be easier to track and document. High-pressure or low-pressure values are reckoned by a system of geometrical weighting procedures.[*]

During July and August 2007 in the U.S., the ridging patterns did emerge and were coincident with a severe drought in the southern Corn Belt and in wide swaths the Southeast. Likewise, a sustained complementary ridge over the eastern Atlantic produced record rains for the U.K. during the predicted period.

However, in a narrow area of the northern U.S. Corn Belt, anomalous rains created flooding conditions during the drought in the Southeast, an unexpected development. As it turned out, the cause of these variations of the predicted fundamental drought pattern was another aspect of planetary motion at work, which was an addition to the eclipse grid influences. In this work, we call this the "Sun line."

The practicality of the eclipse grid is that it allows "geodetic equivalency," the projection of a planet's position onto specific positions on the Earth. Using this concept in his work in the U.K., Jonathan had found that, as the Moon moved past the longitude of the U.K., there was a tendency for troughs to

---

[*]   These, too, are described in articles at docweather.com. See planetary approach and retreat.

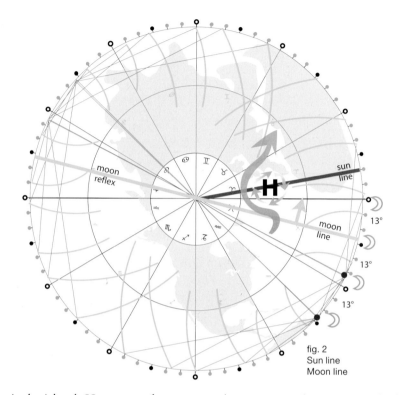

fig. 2
Sun line
Moon line

transit the island. He spent a few seasons documenting the patterns of what he called the "Moon line." I had seen a similar pattern for the U.S. West Coast, so we shared notes on how it unfolded, but it was not something we could use for regular forecasts because it was active only once each month as the Moon passed the position in longitude of the area being studied (see doc-weather.com). Then Jonathan noticed that there seemed also to be an effect whereby the opposite position to the Moon swept the longitude being studied. We called this the "lunar reflex line." With this variation, twice each month we could easily track an apparent lunar influence.

Then another observation by Jonathan moved the Moon line work to another level of predictive value. He noticed that another line could be drawn from the longitude of the Sun to the North Pole on any given day, and that this Sun line appeared to be a position where blocking ridges would form when it passed across the landmasses in the longitude in which it occurred. He further documented case studies showing that, when the lunar line or the lunar reflex line approached a Sun line near the U.K., the result was almost always an enhanced ridge formation that formed on the Sun line as the Moon line or the reflex Moon line approached the Sun line from the west and then passed it to the east. The Moon moves in daily increments of 13° of arc. This means that, in general, the atmosphere in a given longitude moves strongly from west to east for the four days before either the Moon line or the lunar reflex line passes that longitude.

At this point, Jonathan shared these ideas with me and I tried to incorporate them into my system of eclipse grids that is based on the concepts of Johannes Kepler and the Pythagorean monochord. The result of the combination of the Sun line data and the eclipse grid data has been a strikingly elegant fit between the observed phenomena of the highly unusual weather patterns over the U.K. and the Midwest in the summer of 2007.

The following studies are meant to point out the general principles of this unusual set of climate anomalies and are not intended to constitute a causal "proof" of these relationships. They are presented as a phenomenological study and a work in progress. The detailed geometries and mathematical sequences of these phenomena have been logged as notes of the experiment and support these ideas in fine detail. However, they do not appear in this paper because their technical nature would likely be of little interest to the casual reader. The time frames of the study run from April 2007 to August 2007, with the strongest focus on the anomalous drought and flood patterns in sections of the American Midwest and the southern U.K. flooding during July and August of 2007. To give a flavor of how the Sun line works, the spring and summer of 2007 will be presented in a series of charts taken from NOAA sources. The charts show the upper atmospheric anomalies at the 500mb level for two-week periods starting in April 2007.

The following chart shows a set of strong high-pressure areas at northern latitudes for the first two weeks of April 2007. The high pressure is represented by the intense red areas in the center of more general areas of yellow. Low-pressure areas are rendered by larger turquoise areas that gradually deepen in the centers through purples to magenta in the deepest troughs. From the chart it can be seen that strong high pressure dominates northern Siberia and the U.K. These areas are close to being 180° of longitude from each other. Superimposed on the upper air chart from NOAA is an orange cross with one of the legs designated with an orange circle. The cross is composed of the

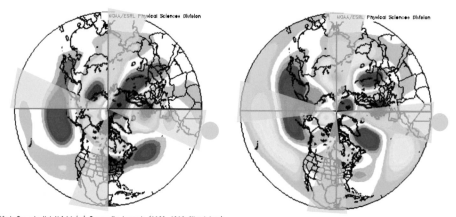

500mb Geopotential Height (m) Composite Anomaly (1968–1996 Climatology)
4/1/07 to 4/15/07
NCEP/NCAR Reanalysis

mb Geopotential Height (m) Composite Anomaly (1968–1996 Climatology)
4/15/07 to 4/30/07
NCEP/NCAR Reanalysis

four projections of Sun lines to form the spokes of a Sun wheel. The orange circle is the position of the Sun when projected onto the Earth using a sidereal placement technique called *geodetic equivalency*. The wedge under this is the position of the Sun for a one-week period. The Sun moves one degree in a day, seven degrees in a week. The wedge defines a zone composed of the area the Sun line would sweep in two weeks. Across the wheel from this wedge is the Sun reflex zone. It is composed of the area the Sun reflex line would sweep in the same period. At right angles to these two zones are the Sun 90 zones. The Sun wheel, with its zones of influence, is projected onto the Earth along with the eclipse grid positions.

It can be seen that during the first two weeks of April 2007 the Sun's position was coincident with a strong ridge formation over the U.K. At the same time another strong ridge formation developed over the opposite longitude in the Sun reflex area. It can be seen that all four Sun zones are areas where high-pressure formations tended to develop. The symmetry of the high-pressure areas in this chart is one example of the many different types of symmetries that this approach allows us to research. An interesting feature of this chart is the placement of major trough formations on the upper air chart during that time frame, which occupy the spaces between the spokes of the Sun wheel. This is a commonly observed phenomenon.

The chart above right shows the following two-week period in April 2007. The Sun wheel has moved east, following the daily longitudinal one degree

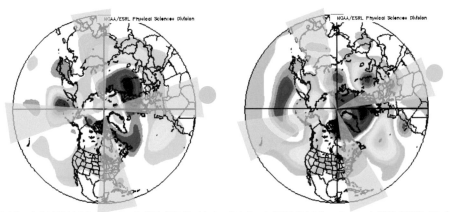

500mb Geopotential Height (m) Composite Anomaly (1968–1996 Climatology) 5/1/07 to 5/15/07 NCEP/NCAR Reanalysis   500mb Geopotential Height (m) Composite Anomaly (1968–1996 Climatology) 5/15/07 to 5/30/07 NCEP/NCAR Reanalysis

increments of the Sun toward the east. There is still a strong upper-level high situated in the area being swept by the Sun line. The reflex position has lost its intensity, and the 90° Sun line over China is the site of a developing ridge. The major low-pressure air masses are still centered between the spokes of the Sun wheel.

In the chart above left, the Sun wheel has progressed 15° to the east and the high-pressure area linked to the solar position has faded. However, the high-pressure air mass linked to the Sun reflex line is very strong over the Aleutian area, and a strong high is linked to the Sun 90° zone over the Midwest. Major troughs are still situated between the spokes of the Sun wheel. In this chart the Sun wheel is once again progressed to the east. Strong high pressure is once again linked to the solar position over Europe, with minor high-pressure areas on the Sun reflex zone and the Sun 90° zone over the northeastern United States. In the chart, a strong mid-latitude ridge is shown over the mid-Atlantic between the spokes, but it is accompanied by the strongest high-latitude low on the chart, situated between Greenland and Iceland. Charts like this show the limits of the Sun line as a total predictive tool. We could ask: What other elements are present that keep the air masses fluctuating as the Sun wheel rotates through the seasons?

The next chart shows a strong high-latitude ridge near the North Pole, sitting exactly on the Sun reflex zone. Major low-pressure areas are once again between the spokes of the Sun wheel.

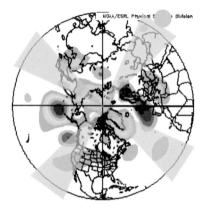

500mb Geopotential Height (m) Composite Anomaly (1968–1996 Climatology)
6/15/07 to 6/30/07
NCEP/NCAR Reanalysis

In this next chart, it appears that the pattern we have been observing has suddenly fallen apart. The strongest high-pressure area is not linked to a spoke of the Sun wheel, but sits over the Aleutians exactly in the space between the Sun 90° zone and the Sun reflex zone. Moreover, all upper air masses are greatly reduced in intensity. This development points to the need for another layer on the data field. Put another way, what else is working in concert with the Sun wheel?

The previous charts were intended to help familiarize the reader with the idea of the Sun wheel and it's various zones. In these research protocols, the Sun wheel influences are integrated into another set of geometric variables known as the *eclipse grid*. The chart shows the eclipse grid pattern for the drought in 1988, which wreaked havoc on U.S. agriculture during that growing season. The key element in the next chart (opposite) is the set of curved lines on the grid centered over the American inter-mountain area. The red curves are known as *jet curves* in this system. Each one is placed at a distance of 45° from one of the eclipse points. There are two 45° jet curves extending westward from the two eclipse points over the western Atlantic. There are also two 45° jet curves (red) extending eastward toward the U.K., from the eclipse points. The placement of the curve at 45° from the eclipse point is known as an *angular aspect*. The curve is at an angular aspect in that it is at a 45° arc angle from the eclipse point. An arc angle reckons the distance in degrees from one point on a circle to another point. Planets whose positions in longitude are placed on the circle are also found at arc angles from the eclipse points and move in rhythmic cadences in predictable time frames. The timing of these motions against the eclipse point is the key to the forecast system. The numerical values that give weight to particular angular aspects are the source of high- or low-pressure values generated on an eclipse point. These numbers are generated by each motion-in-arc event from each aspecting planet. The

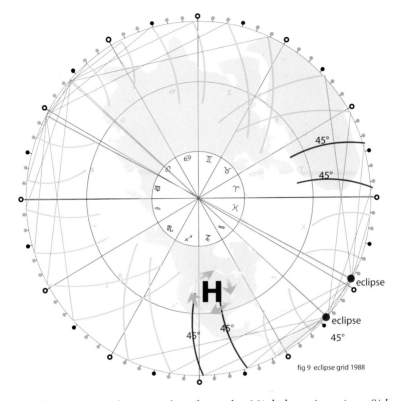

fig 9 eclipse grid 1988

raw data for these numbers is taken from the *Michelsen American Sidereal Ephemeris*, 2001 to 2025.

In these protocols, when a planet forms an angular aspect against one or another of the eclipse points, the latitude and longitude of the projected jet curves have proved to be reliable, statistically significant areas for predicting the emergence of anomalies in the long-wave oscillations of the jet stream. That is why these curves are called "jet curves." The jet curve can be the expected location of either a high-pressure blocking area or a low-pressure trough, depending on the numerical value generated by a given motion-in-arc event.

It can be seen in the chart that the jet curves over the western U.S. bracket the position of an anomalous high-pressure area. That area, forming in April 1988, was the source of the devastating drought that started in the High Plains. Through May and June of 1988, the blocking ridge expanded eastward to

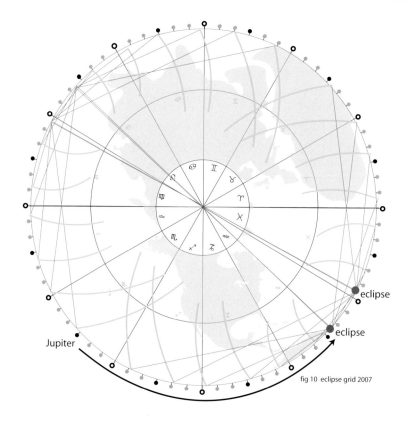

fig 10 eclipse grid 2007

settle over the Great Lakes, wiping out the corn and soy crops that year by blocking intrusions of the storm-bearing jet stream.

This next chart is the eclipse grid chart for summer 2007. We can readily see that the placement of the eclipse points and the subsequent placement of the 45° jet curves from these points are very similar to those in the 1988 chart. In 1988, an angular aspect between Jupiter and the eclipse points in the early spring kept high-pressure values on the two eclipse points for weeks at a time. That extended relationship was the underlying stagnant atmospheric signature of the drought period. In 2007, there was a repeat of that extended high-pressure relationship between Jupiter and the eclipse points. Jupiter stood at a very significant numerical angle in relation to the western eclipse point. In the past, that angle has been coincident with sustained periods of blocking. The difference between the two years was the stagnant period in 2007 that occurred in July and August instead of April, May, and June. Since the

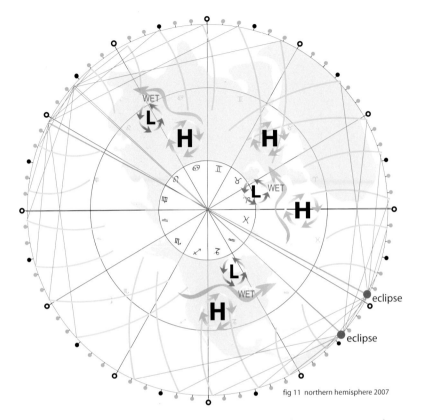

fig 11 northern hemisphere 2007

drought in 2007 turned out to be less extensive than the one in 1988, the question is: Why did a similar pattern in 2007 not produce a reaction as extensive as the drought of 1988? To try to answer this it helps to weave the eclipse grid data into the Sun-wheel data. With this weaving, the unusual events of the summer of 2007 can come into view.

The anomalous flood–drought pattern in the Corn Belt during the summer of 2007 was played out in several widely separated geographical areas. In the eastern Atlantic, there was a pattern in which the U.K. received flood conditions, while Europe was caught in a drought. In eastern Asia, on the opposite side of the Earth from the Midwestern U.S., Southern China was the site of tremendous flooding, while neighboring states in Northern China endured crippling drought under a persistent blocking regime. In the U.S., twenty counties in the Northern Corn belt were the site of highly unusual flooding, while in neighboring counties and across the Southern Corn Belt the

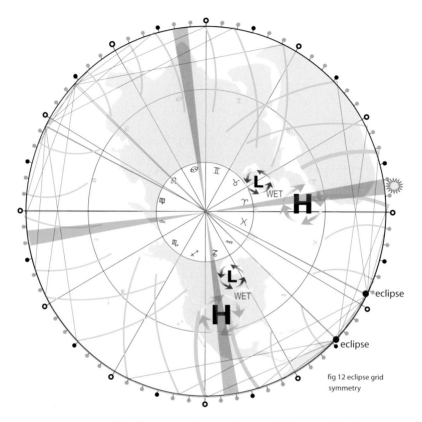

fig 12 eclipse grid
symmetry

summer was memorable for severed drought conditions, with Georgia and Tennessee having century-long droughts. In each case, a stubborn, persistent ridge over the drought area created a pattern in which the long-wave westerlies (actual trough patterns in the jet stream) could not break the ridges, but moved around them to deposit copious rains in specific areas while the high pressure remained intact. The unusual conditions in each remained while Jupiter and the lunar node were at high-pressure values in relation to the eclipse points.

In the U.K. and the U.S., the position of the eclipse grid involved the placement of 45° jet curves (blue), which generally support mid-latitude influences. In this system, the 45° jet curves have frequently proved to be the sites of blocking patterns. In July and August 2007, the prevailing high pressure from Jupiter and the lunar node blocked ridges in place as the Sun wheel (pink cross) transited each pair of 45° jet curves. This observation was

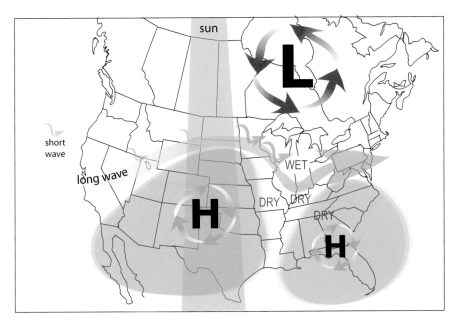

the product of a process of intensive, event-by-event tracking protocols of planetary motion-in-arc events every day during the summer.

As pointed out, the original idea for the study was that the placement of the sets of 45° jet curves from the points in the western Atlantic would be more easily tracked owing to the symmetry of 45° jet curves extending from the eclipse points in the central Atlantic. The sets of 45° jet curves are reciprocally harmonic to each other on opposite sides of the eclipse points. In 2007, that symmetry put sets of complementary 45° jet curves over the High Plains of the U.S. and off the west coast of the U.K. That made 2007 an ideal study year, especially with Jupiter and the lunar node at angular aspects that sustained high pressure on the jet curves for a six weeks.

As it turned out, as the Sun wheel moved east through July and August, the ridges linked to the eclipse grid also migrated slowly eastward from one 45° jet curve to the next. That was the case in the U.S. Corn Belt and in the U.K. The long-wave jet-stream patterns (Rossby waves) linked to the Sun wheel were the sites of the persistent high-pressure areas that formed along successive jet curves of the eclipse grid (pink). The combination of the curve and the Sun line dominated the atmosphere in those particular longitudes.

This produced a strong polarization of the storm track in the Midwest in July and August.

In the U.S., the result was the weak, short-wave frontal passages at the mesoscale level (thunderstorms and squall lines), which just kept rolling along the persistent ridge over the southern high plains and the southern Corn Belt and then dropping into the wet area in northeastern Iowa and northern Illinois with wave after wave of thunderstorms. The Atlantic points of the eclipse grid were under the influence of high pressure from Jupiter and the lunar node for about two months, with only brief shifts to low pressure on the eclipse points. That kept the established ridges strongly in place. At the same time, the Sun wheel's motions from west to east during those two months shifted the position of the ridge patterning from one 45° jet curve to the next. As that happened, the ridges migrated from west to east across the U.S., though not far enough to destabilize the pattern owing to a particular configuration of the eclipse grid. The following sequences show this unfolding of unusual summer weather events across the Corn Belt. Later, we will refer to the symmetrical events in the U.K.

The adaptation of a government chart shows the upper air pattern for July 1, 2007, when Jupiter was at a strong high-pressure angular aspect to the eclipse points. The blue jet curves show this influence. Torrential rains in Texas and Missouri arose then because of circulation around a persistent high centered on the western 45° jet curve and spread over the Rockies. This was accompanied by a persistent high over Florida on the 22° jet curve. Between the two, a gap formed in which a low-pressure area became trapped for days on end. The resulting monsoon pattern brought abundant moisture from the Gulf of Mexico into the southern High Plains and created devastating floods in Texas. The key was that the trough could not go anywhere because the ridges were too strong, but not strong enough to link to each other.

In this system, Jupiter was instrumental in very strong blocking high patterns. With Jupiter at a high-pressure angular aspect to an eclipse point, blocking situations tend to persist. That persistence figures strongly in the Texas flood scenario. The ridge that resulted from the Jupiter aspect on the 45° points blocked the western third of the U.S. from Nevada to the High Plains east of Denver. The trough was caught in the lee of the blocking ridge. The jet curves are superimposed on this government upper air chart to show

070701/0600V012 NAM 500 MB HGT, GEO ABS VORTICITY

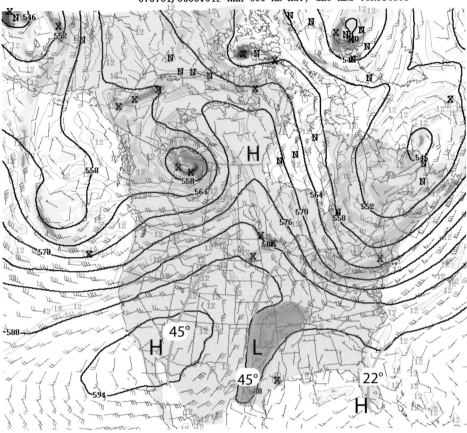

the similarities between the jet-curve patterns and the actual air masses that resulted from them.

The next chart (following page) shows the whole northern hemisphere during these events. It shows the extent to which the high-pressure areas were generated by Jupiter as it formed high-pressure angular aspects against the eclipse points. Every blue line is a high-pressure value from the eclipse points. All of the jet curves show high pressure. The high-latitude lows conform to the flow pattern of the jet curves.

In the Gulf of Alaska, the low there is in the lee of the high-latitude jet curves extending toward the Bering Sea. It stayed there for weeks. In the North Atlantic the high-latitude lows there are integrated into a jet stream streamline pattern extending from the U.S. East Coast to the U.K. This

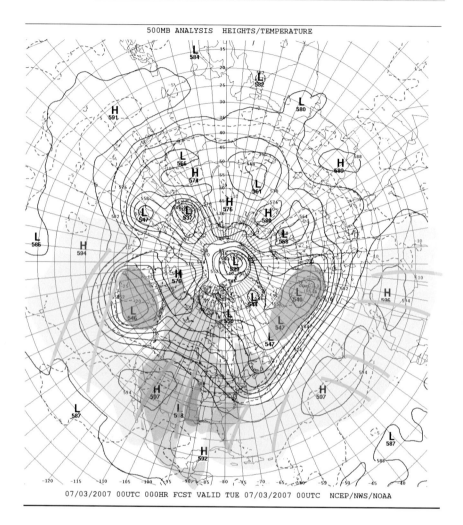

500MB ANALYSIS HEIGHTS/TEMPERATURE

07/03/2007 00UTC 000HR FCST VALID TUE 07/03/2007 00UTC NCEP/NWS/NOAA

streamline rides on the crests of the four 22° jet curves, spanning the Atlantic from Florida to the Cape Verde Islands. This spanning stream of zonal flow is broken when it reaches the ridge formed by the two 45° jet curves over western Africa. Typically, a zonal jet from one end of the Atlantic to the other will always curve toward the north as it approaches the continent. Also typically, a zonal jet such as this is supported by a set of jet curves that are all aspected to high-pressure values. These sets of curves provided the impetus for the arrow-like jet stream that brought inundating rains to the U.K. during spring and early summer 2007.

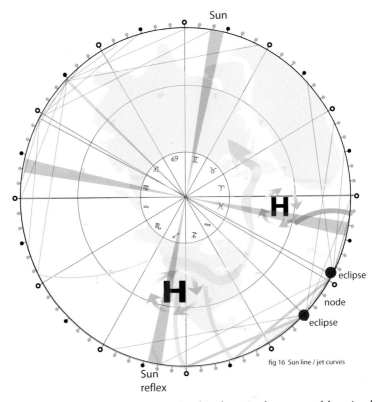

fig 16 Sun line / jet curves

The only other low-pressure area in the chart is the trapped low in the lee of the two 45° jet curves over the western U.S. This constituted the unusual nature of this storm pattern for the southern Plains. There was no reason why the low should have been so persistent, except that it was caught between remarkably persistent areas of high pressure.

As the pattern between Jupiter and the node settled into place, most of the area between the mid-Pacific and the Sahara desert was given over to high pressure at mid-latitudes. Low-pressure areas rotated around the Arctic Circle like a high-latitude crown, but seldom did a large loop of the jet stream form in a deep trough where there were jet curves. However, over Texas the strange vertical lee trough between the strong western ridge over the Rockies and the high over the Southeast was very persistent.

During this period in 2007, it was expected that the ridges over the U.S. would eventually merge into a single ridge at mid-latitudes, which would signal the beginning of a very dry period for the whole of the High Plains

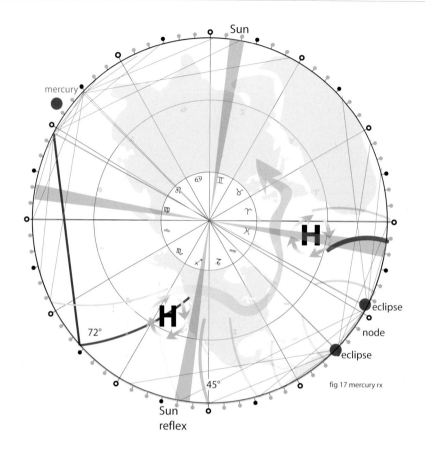

fig 17 mercury rx

and Corn Belt. The dominance of Jupiter at these particular configurations of high pressure went from June 23 to July 6. During that time, almost everywhere in the U.S. was under a persistent regime of high pressure. During that time, however, Texas, Oklahoma, Missouri, southeastern Iowa, and eastern Nebraska received record rains as the weak low brought up waves of moist air from the Gulf of Mexico. This moisture interacted with fronts from the northwest to create lines of thunderstorms. Why did the high over the Rockies not link to the high over the Southeast? There seemed to be no answer, and the phenomena continued to repeat. However, by July 6 Jupiter had reached the term limits of its tenure. On that day, the lunar node shifted to high pressure on both of the eclipse points in the western Atlantic.

In the course of this study, it became clear that there was a relationship between the movements of the Sun wheel and the placement of the eclipse grid with its jet curves. This became observable in the beginning of July as the Sun reflex line began crossing the West Coast. On the first of July 2007, the Sun was in the longitude of the east coast of Africa. That put the Sun reflex line through Vancouver Island in the Pacific Northwest (pink zone on the West Coast). In that position, it approached the western 45° jet curve that had just had a motion-in-arc event from the lunar node (thick blue curve). That jet curve was aspected to a high-pressure value by the node. The Sun reflex zone touched that newly aspected jet curve, and a strong ridge built over the West Coast of the U.S. In the U.K., a ridge was also placed near the Sun 90° line off the coast of Africa, where it crossed the 22° jet curve from the eastern eclipse point. This high was related to a nodal shift on the first of July to high pressure on the eastern point of the Atlantic pair of eclipse points. It formed where the 22° jet curve touched the Sun line.

Then, on July 8, Mercury (red disc over Asia) went on station at a high-pressure value to the 72° jet curve (red curve) crossing just north of Vancouver Island. The Sun reflex position line had moved east with the Sun and was over the Nevada basin at the time. Through the end of June, the center of the coast ridge following the Sun had been moving slowly east toward the 45° jet curve over Utah (green curve). On the day of the Mercury station, the ridge stopped following the Sun to the east. It disappeared that day and reappeared the next day to the west in a position directly on the 72° jet curve in the ocean off the West Coast. In other words, the ridge moved from the 45° jet curve (green) that the Sun line was approaching in the high desert to the more recently active jet curve (red) that was just aspected by the Mercury station event. The Sun line at that time was crossing the West Coast and was in between the 72° curve over Vancouver (red) and the 45° jet curve over Utah (green). This sudden shift toward the most recently active jet curve showed that the most recently active jet curve in the vicinity of a Sun line is where a block is most likely to form in a given situation.

To substantiate this idea, on the day that Mercury went off station (July 13), the ridge began to fade over Vancouver and on July 14, as Jupiter aspected the eastern points at high-pressure values, the block again shifted to the 45° jet curve over Utah. On that day, the ridge moved to the east toward this new, more active jet curve as the Sun line also approached this same jet curve. With Jupiter at high-pressure values to both eclipse points, extreme high pressure

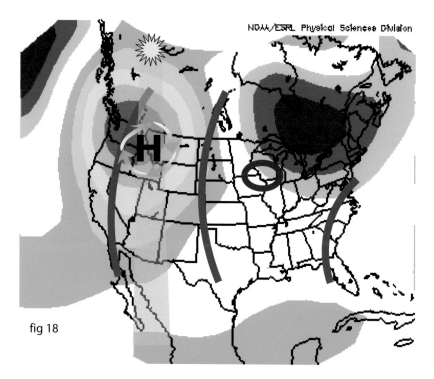

fig 18

500mb Geopotential Height (m) Composite Anomaly (1968–1996 Climatology)
7/1/07 to 7/15/07
NCEP/NCAR Reanalysis

dominated the upper air charts between Hawaii and the U.K. The strongest highs, however, were linked to the active jet curves touching the spokes of the Sun wheel.

On July 15, 2007, the lunar node shifted to a high-pressure angular aspect to the eastern 45° jet curve on the Continental Divide. It was thought that this shift would create a strong ridge formation in the middle of the continent, keeping the summer rains from moving into the northern Corn Belt. In 1988, an analog year for 2007, a ridge in this position in May and June had created an extreme drought that wreaked havoc on Corn Belt crops that year. The Corn Belt is divided among western, northern, and southern states. The western states are northern Oklahoma, Kansas, and Nebraska. The northern states are the Dakotas, Minnesota, Wisconsin, and Michigan. The southern states are Missouri, southern Illinois, southern Indiana, and southern Iowa. These are the regions where corn and soybeans are grown. However, the vast

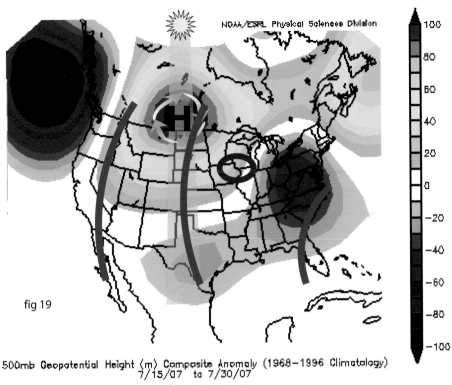

500mb Geopotential Height (m) Composite Anomaly (1968–1996 Climatology)
7/15/07 to 7/30/07
NCEP/NCAR Reanalysis

fig 19

majority of soybeans that make up the soy market are grown in a twenty-county area (red oval) centered on northern Iowa, southern Minnesota, northern Illinois, and northern Indiana. This area is the prime soy-producing area of the continent, since it is the area most likely to receive beneficial rains even during dry growing seasons. This occurs because, during the summer months, a cold prevailing summer storm track from the Canadian plains meets a warm prevailing summer circulation of moist air coming north from the Gulf of Mexico. The meeting of these two streams provides the moisture needed by the grain markets in the United States and the world.

The shift of the lunar node in mid-July signaled a radical change in the patterns for the northern hemisphere jet stream in both the U.S. and the U.K. In the United States, a ridge pattern developed on the Sun line in which very weak short waves of cold began tracking over the continental ridge. These were guided by a strong low over Hudson Bay that is not shown, since it is

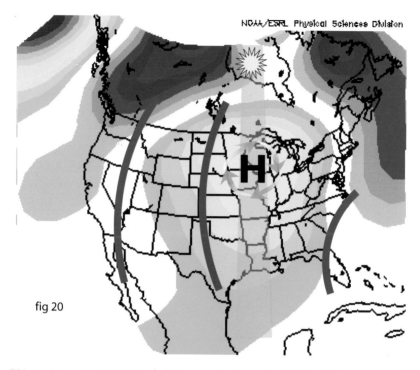

fig 20

500mb Geopotential Height (m) Composite Anomaly (1968–1996 Climatology)
7/30/07 to 8/14/07
NCEP/NCAR Reanalysis

north of the chart area. The fronts, or short waves, transited the ridge over the High Plains and then dropped along the eastern border, right into the very heart of the Corn Belt (red oval). Once there, they deposited copious rains in a small but very important area in the scheme of things for the soybean market.

A short wave is a smaller ripple in the jet stream that moves through the large loops of the jet. Short waves bring frontal energies that produce very local thunderstorms and front lines in specific areas. The short waves in this scenario rolled over the persistent ridge depositing highly unusual rains in the twenty county area of the central Corn Belt. The area of the copious rains was in northeastern Iowa, northwestern Illinois, and southeastern Minnesota. This prime soybean-growing area extends to about twenty counties and was the site of many such events in the middle of July. In each rain event, the same

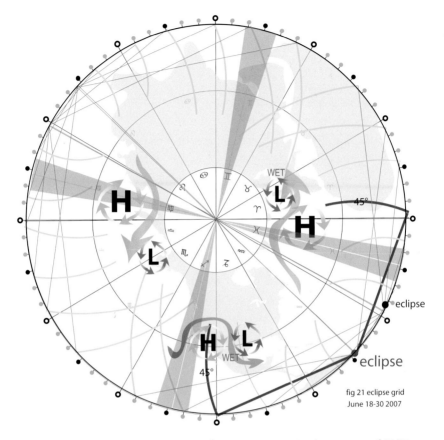

Symmetry between U.S. air masses and U.K. air masses in the summer of 2007

areas received ample rains, whereas much of the rest of the Corn Belt sat in drought conditions.

In early August, the ridge followed the Sun line east into the Corn Belt and broke the rain pattern. Then, as the lunar node moved off its angular aspect in late August, the pattern diminished and the ridges broke down. The rains stopped in Iowa and Chicago, and the dryness was tempered in Missouri and southern Illinois. At the height of the pattern, one county in northern Illinois registered twenty inches of rain and another two hundred miles to the south registered .2 inches of rain, while large areas of the southern Corn Belt and Mississippi Valley were the sites of intense drought.

The unusual flood-and-drought pattern that developed in the Corn Belt in the U.S. during the summer of 2007 was also played out in the U.K. In late

June 2007, a persistent ridge formed over the Rockies in the U.S., where the reflex Sun line was passing over the eclipse grid. Downstream, to the east of that ridge, a persistent trough produced record flooding over Texas. At the same time, over the Atlantic on the 90° spoke of the Sun wheel, a persistent ridge formed that was the source of record flooding downstream of the ridge in the U.K. A third ridge formed on the opposite Sun 90° line over the Bering Sea, with a strong trough in the lee of that ridge. Three symmetrical air masses maintained these ridge-and-lee trough patterns in the same time frames.

The U.K. and U.S. patterns were formed on the Sun line spokes closest to the 45° jet curves from the eclipse points in the western Atlantic. The eclipse point and the associated complex of lines and curves are rendered in red. That eclipse point was stimulated to a strong high-pressure value by a strong aspect from Jupiter on the westernmost point of the pair. We can see that, where these persistent high-pressure curves originating on the western Atlantic eclipse point came near to a spoke of the Sun wheel, the resulting air mass was a strong high-pressure area. These types of ridge formations are *omegas*. In an omega pattern, the ridge stays in place and extends latitudinally to a polar position. On either side of the latitudinally extended ridge, strong and persistent lows form at a lower latitude. In the summer of 2007, the lee areas of these omega ridges were the sites of anomalous flooding. These events were symmetrical in both time and space. In the June flooding sequences, the omega ridges dissolved into a more general zonal, or horizontal, flow the very day that the lunar node moved in arc deleting the strong Jupiter influence on the eclipse point.

In the first two weeks of July, both the Atlantic and the central U.S. became more settled. The blocking ridges creating the lee trough patterns diminished in the same time frame. The shift in conditions occurred when the lunar node moved to a position that supported more generalized ridge formation rather than omega ridge formation. The flooding in the U.K. abated, and flood conditions in Texas eased, as well.

On July 15, a strong omega pattern unfolded once again on both the High Plains of the U.S. and in the mid-Atlantic. The lunar node had moved in arc to a strong omega, forming position in relation to the eastern eclipse point. This position is shown in red, along with the associated lines and jet curves. In this scenario, it was the 22° jet curves over the Atlantic that supported the omega

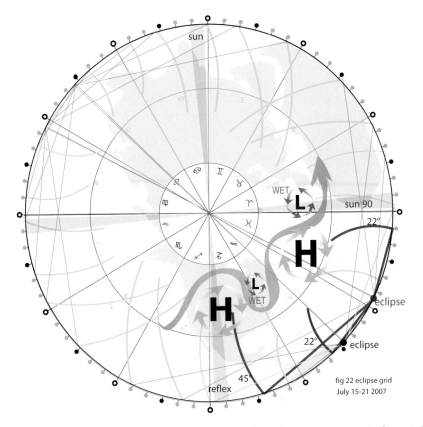

fig 22 eclipse grid
July 15-21 2007

for the U.K. The position of the Sun lines at that time was not so influential in the positioning of the mid-Atlantic omega block, since the block formed to the west of the Sun 90° line.

However, at the same time, the Corn Belt of Iowa and Chicago was the site of a persistent lee trough downstream of a blocking ridge over the western High Plains of the U.S. This blocking omega ridge was centered on the site where the 45° jet curve from the eastern eclipse point crossed the Solar reflex line over the Rockies. This omega created anomalous flooding conditions in the Corn Belt that were synchronous to the events in the U.K. Using the eclipse grid, we see that the events in Europe and the U.S. were once again symmetrical in time and space. In both the U.S. and the U.K., blocking patterns, a stubborn ridge prevented the long-wave westerlies from dislodging the block, so the storm track just kept rippling over the ridge top and diving into the wet area with pulse after pulse of short-wave storms.

From a planetary point of view, the eclipse grid was under the influence of general high-pressure angular aspects from Jupiter and the lunar node for about two months, with only brief shifts to low pressure on the eclipse points. That kept the established ridges strongly in place. The Sun wheel motions from west to east through those two months shifted the ridge patterning from one jet curve to the next. As that happened, the ridges migrated from west to east, but not far enough to shake the pattern, since the particular configuration of the eclipse grid in both places involved the symmetrical placement of 45° and 22° jet curves, which generally support mid-latitude influences and are prone to persistent blocking when aspected to high-pressure values.

In general, as the Sun wheel moved east through July and August, the blocking ridges also migrated slowly eastward from one jet curve to the next. Since it took the Sun wheel six weeks to migrate from one side of the U.S. to the other, and since those curves were under the influence of high pressure for most of that time, that gave rise to blocking ridges that stayed put for quite a long time. That was the case synchronously in the U.S. Corn Belt and in the U.K. The long-wave patterns (Rossby waves) linked to the Sun wheel encountered persistent high pressure along the eclipse grid, and the ridges polarized the atmosphere. The result was a strong polarization of the storm track in those areas. The polarization resulted in the formation of persistent lee troughs that produced the rains in the Corn Belt and in the U.K. The troughs formed in the lee of the prevailing ridges as short waves moved across them and brought numerous and repetitive frontal passages resulting in flood conditions in both areas. The transiting lows in the lee troughs of the blocking omega ridges were the sources of the anomalous rains in the U.K. and the U.S. during July and early August 2007.

CHAPTER 11

DECADAL WEST COAST SUMMER EVENTS

This chapter arose from research by Climatrends (climatrends.com) long-range forecast service, which produces annual extended forecasts for Sonoma County wineries. A critical issue for the California wine industry is a tendency there for late heat waves to arise in the summer, in July, August, and September. During certain years, late summer heat may coincide with *veraison*, the onset of the grape-ripening process. When grapes are in the ripening phase, they are very susceptible to extremes of heat. A late heat spike in excess of 100° Fahrenheit (38° Celsius) for only a few days in August or September can devastate a promising harvest, with serious financial repercussions. If there is ample warning of a heat spike, growers can irrigate and save their crops. If the heat never arrives, however, irrigation of the vineyards can compromise grape quality and, during late veraison, the skins become thin and the extra water may cause the grapes to burst, with the result that fungus spreads quickly into the sugary sap of the split-open fruit.

Thankfully, a unique climate pattern on the West Coast makes the likelihood of a devastating heat spike during veraison merely a probability rather than a certainty. While temperatures in the California Central Valley may exceed triple digits several times every summer, temperatures in excess of 100° are only a rhythmic phenomenon at best in the coastal mountains and grape growing areas. The reason for temperate conditions on the Coast is that the jet stream is pushed far to the north in the Gulf of Alaska each summer by the growth of the Hawaii high, whereas regular breakouts of low-pressure cells drift south and drop out of the west-to-east flow of the jet stream. The low-pressure cells then idle off the coast. Climatologists call this phenomenon a *cutoff low pattern*. The lows are cutoff from the impetus of the jet stream, and since there is little west-to-east movement, they tend to hover off the coast and send cooling winds into the wine country when the scorch is on in the valley. This microclimate is the reason California wines are a major agricultural activity for the state. Without cutoff lows, most wine grapes would become raisins almost every year.

Predicting the precise influence of cutoff lows drifting around outside the jet-stream circulation is risky business for a long-range forecaster when there is a lot of money at stake, which is the situation for wine producers. Over time, Climatrends provided several busted extreme-heat forecasts, which prompted a survey of major heat events for Sonoma County. The wine areas surrounding two cities were chosen for the study: Napa and Glen Ellen. Napa sits on the floor of a broad valley about forty miles from the coast, whereas Glen Ellen lies in a narrow mountain valley that has access to coastal fogs when the marine layer is strong, but is also situated fifteen miles inland in a "banana belt" that can sustain serious heat when conditions are right.

For the survey, the record-high temperatures in July and August for each location were used to set up thirty-five case studies. In each case, record-high temperatures were chosen that lasted two or more days. The study ignored single days of record-high temperatures, because there are so many variables, even in a cool year, that a strong low moving into an established high over the coast can pressurize it to the degree that a temperature spike results.

First, the study noted that years during which heat was a serious issue seemed to be grouped into ten-year (decadal) periods. This reflected a commonly observed climatic phenomenon known as "decadal influences," which is the tendency for extreme events in the atmosphere to occur at approximately ten-year intervals. The decadal influence is close to the periodicity of the lunar node moving backward through the zodiac in a 9.3-year rhythm. Using the eclipse grid as a basis for the study, researchers could see that the phenomena of cutoff lows and heat spikes show a remarkable coincidence with the motions of the eclipse grid.

To illustrate this principle, charts from National Oceanic and Atmospheric Association and the Climatic Data Center will illustrate the following survey. The eclipse grid pattern present during July and August of the selected years overlay the government charts.

The year 1991 saw a very cool August on the Coast. In the eclipse grid, 45° jet curves straddled Hawaii and were the source of a high-pressure area. This pattern was the initial entry of the bottom of the basin into the eastern Pacific, because two of the four 22° jet curves were off the Coast and represented a

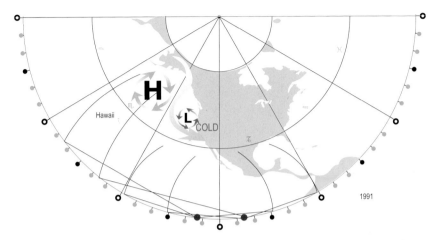

low-latitude influence. In the chart above, the area of these curves was the site of an anomalous low-pressure trough coming from the Aleutians. It lasted for most of the month and created cool conditions on the West Coast.

This chart from NOAA depicts the anomalous highs and lows present in August 1991. The chart is over-laid with a modified eclipse grid that shows the essential elements of the eclipse grid projection for that period. The purpose is to integrate 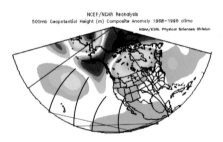 the climate data of government sources with the eclipse grid. We see that the anomalous high over Hawaii and the low over the Aleutians create a situation similar to that depicted in the first chart. The following series of charts will be similar. They are intended to show how the migration of the eclipse grid through decadal patterns is coincident with observable climate sequences.

1992: Cool weather in Southern California continued with moderately cool weather on the central coast, as low pressure at the bottom of the basin pattern dominated the month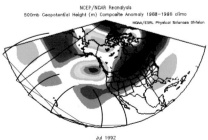

of July. The low centered on the four 22° jet curves in the eastern Pacific. Persistent high pressure off western Canada drew a dominant cutoff low into the basin.

1993: The cool summer trend continued in 1993, as all 22° jet curves were over the ocean in a pattern that was truly the bottom of the basin. A strong ridge in the central Gulf of Alaska created a downstream low, centered on the two 45° jet curves over Idaho. These two curves represented the beginning of a transition to the uphill, or block side, of the crest in the eclipse grid.

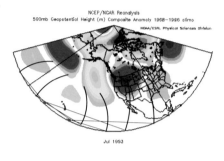

1994: During the summer of 1994, temperate conditions emerged as the basin pattern moved west to settle over Hawaii. The block side of the crest put the two 45° jet curves astride the Coast. A moderate low off Oregon joined with a moderate ridge over Idaho and kept conditions temperate for this summer. This pattern is an analog to the year 2003.

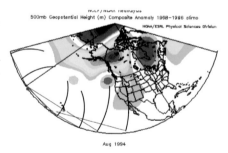

1995: The general West Coast circulation for 1995 was a shift to temperate conditions. A weak ridge in the Gulf of Alaska kept the weather mild. A cutoff low due west of the Golden Gate was trapped between the crest of the block side with the eclipse diamond over the Rockies. The cutoff low created an

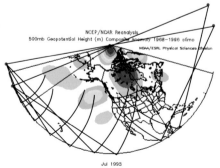

unusual, isolated cold pocket from San Francisco to Lake Tahoe, while everywhere else had above-average warmth.

1996: The eclipse diamond centered on the Nevada–California border and produced a scorcher of a summer on the Coast. A persistent ridge locked onto the diamond, and the height of the crest of the eclipse grid kept the West Coast trapped under high pressure from San Jose to Seattle.

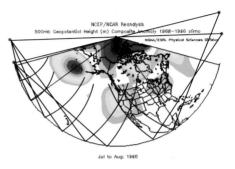

1997: Although the eclipse grid moved west with the motion of the eclipses, the diamond was situated just off the Coast during August, when a low trapped in the center of the Gulf of Alaska kept a ridge formation over the West Coast. Several heat events accompanied the ridge. However, the grid had progressed

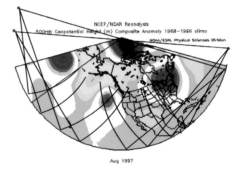

to the cascade side of the eclipse crest. Next up were the two 45° jet curves on the cascade side of the grid.

1998: During the summer of 1998, the two 45° jet curves on the cascade side of the eclipse grid crest centered over Idaho. A persistent ridge formed there, bringing heat to the West Coast. The ridge was fed by a very strong low over the Aleutians that centered on the crux of the eclipse diamond. This was a classic pattern, ending a cycle of

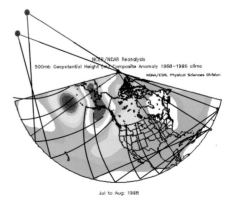

five years of warm conditions as the crest of the grid passed across the coast. The next cycle turned to cool as the bottom of the basin in the grid began to transit the Coast.

1999: The grid had progressed the diamond out over the central Pacific, and the two 45° jet curves were over the central Gulf of Alaska. This created a space for lows to cutoff from the general circulation and drift into the Coast to keep the coastal regions abnormally cool. The grid had moved to the bottom of the basin as lows cascaded from the Gulf of Alaska into the Coast, where they became cutoff from the general circulation.

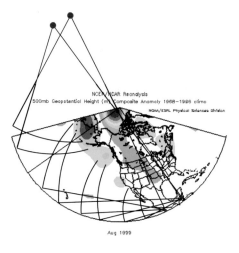

2000: The two 45° jet curves straddled Hawaii, and a strong ridge locked onto them in the lee of the Aleutians. This created a space where lows could cutoff and drift into the West Coast, keeping the weather abnormally cool on the Coast.

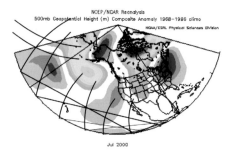

2001: The basin pattern moved farther west into the East Pacific as a strong ridge over the Aleutians kept the potential for cutoff lows being trapped against the coast. This again created cool and temperate summer conditions for the West Coast.

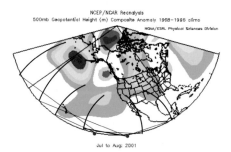

2002: The whole basin pattern was in the eastern Pacific in August 2002. A ridge lodged against the Canadian West Coast, and the north-to-south circulation of the exit jet of the ridge brought cool air down from Canada and drew low pressure from the

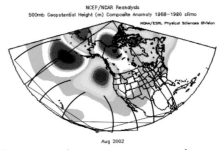

Hudson Bay area into Idaho. The result was another temperate-to-cool summer in the bottom of the basin pattern. The passing of the 22° jet curves into the ocean signaled a change since the next set of curves would be the beginning of a block pattern once again as the two 45° jet curves of the block side of the eclipse crest moved westward from the Plains states. In the block pattern, the prevailing jet stream vector is south-to-north, or ridge-forming.

2003: This was a classic scorch pattern for the West Coast, as the two 45° jet curves on the berm side of the crest centered over Idaho and Montana. The circulation around the ridge brought a ridge of high pressure over the desert and hot conditions for

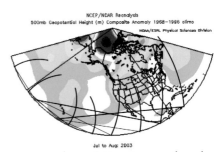

the West Coast. This grid focused in a more easterly position compared to the analog of 1994. Consequently, the ridge focused more in an easterly position. The resulting weather on the Coast was hot compared to temperate conditions in 1994, when the ridge centered on the two 45° jet curves was in British Columbia, and circulation brought coolness down into the West Coast rather than the circulation in 2003 that created continental heat.

2004: Heat once again dominated in July 2004, as the eclipse grid diamond fixed over Denver. A persistent ridge formed on the block side of the eclipse crest with a blocking ridge forming over western Canada. That circulation kept the marine

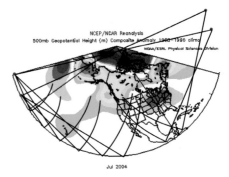

layer at bay on the West Coast, making for hot conditions during most of the month. Cutoff lows formed when the ridge moved to the Aleutians instead of forming on the continent.

2005: The heat on the West Coast was highest in July and August 2005. The eclipse grid crest centered on the desert, and a persistent blocking ridge formed over the northwestern portions of the U.S. Circulation from the ridge brought continental heat into the western states and to the coastal areas of California.

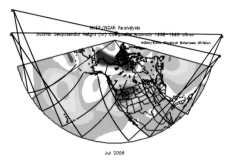

2006: A strong pattern of heat on the West Coast emerged in July 2006. The eclipse diamond was positioned just off the Coast, where a strong blocking ridge over Idaho established a continental flow into California, raising temperatures there and across the whole Northwest. This season rep-

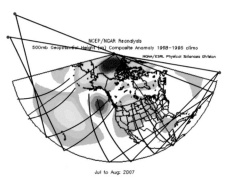

resented the center of the transit of the eclipse grid crest. The next eclipses would move the peak out into the Pacific and introduce the north-to-south flows of the cascade side of the crest.

2007: The eclipse grid crest has moved to the center of the Gulf of Alaska. The pair of 45° jet curves (the next set of curves in the cascade side of the crest) were situated over the Midwest. In July and August, a persistent ridge formed there over Montana, bringing a continental flow to the western states. In California, the central valley was a site of strong heating but the coast remained temperate owing to a persistent low in the Gulf of Alaska, which kept the marine layer active. That low centered on the eclipse diamond.

2008: A pattern similar to that in 2007 emerged with heat in the central valley of California and temperate conditions on the Coast. The eclipse diamond was over the western Gulf of Alaska with a strong, persistent low that steered a trans-Pacific jet into the West Coast, creating a ridge over Northern California and Nevada. The result was heat in the California Central Valley and temperate conditions on the Coast,

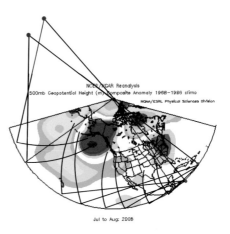

Jul to Aug: 2008

as the marine layer was active in putting weak cutoff surface lows off the coast when the valley heated up.

The final chart (next page) shows a composite of periods when heat spikes in California wine country were the most severe since 1944. We can see that the most dangerous periods by far for late heat spikes in summer occurred when the eclipse diamond was either straddling the Coast or over the eastern Pacific. In that position, the 72° jet curves extend into high latitudes over the Western States. Ridges that form when the grid is in such a position tend to be very stable and surge to high latitudes when the Sun line supports the grid. A strong ridge on the continent when the eclipse diamond is near the coast tends to center the ridge over Idaho and Montana. A ridge in that position brings strong warmth across the Sierras and down into the valley. When it is strong and stable, such a ridge pushes the marine layer to the west and the scorch is on for the wine country. The probability for heat is greatest at these times. It is possible for heat to arise at other times, as well, but with much less probability. It is also possible to have the grid configuration with no significant planetary involvement. In that case, there is no heat spike.

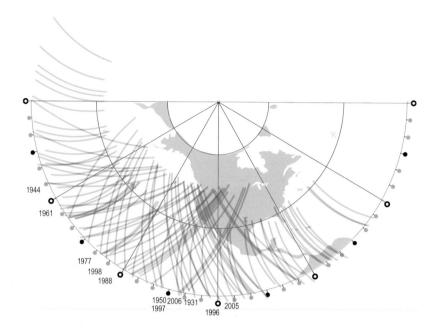

Each year, when doing the background research for producing a Climatrends extended forecast for Sonoma County, the eclipse grid motions have been a great aid in understanding these subtle climatic patterns that support this major agricultural industry of the State of California.

CHAPTER 12

MORAL ROOTS OF THE CLIMATE CRISIS

In his widely acclaimed movie *An Inconvenient Truth,* Al Gore concludes that the climate crisis is actually a moral crisis. Exactly what does that mean? How does morality—generally considered a soul or spiritual value—relate to the science of climate research? Looking for an analog or image to answer this question, the legend of Cain and Abel might provide a beginning.

In the legend, Abel was the first son of Adam and Eve. He became what contemporary theorists would call a *forager.* He did not produce food, but gathered what was available. We might call Cain, on the other hand, a *pastoralist,* or *agricultural technician.* He designed instruments to prune trees, to plow the soil, and to control fire. The ancient god Jehovah recognized the sacrifice of Abel, because Abel did not attempt to transform the given natural world. Jehovah did not recognize Cain's sacrifice, because the fruits and vegetables presented for sacrifice were not taken directly from nature but were "engineered" by Cain in his own way. This difference is the root of the fratricide that lies at the beginning of technology. The pastoralists were not only adept at making tools, but they were frequently bellicose, as well. The most significant and rapidly developed technologies in the shift from Paleolithic times to the Neolithic revolution developed around the forms of increasingly sophisticated arrowheads. That should sound distressingly contemporary, but it is also the central picture in most ancient myths.

An aggressive urge to dominate and subdue the natural world often accompanied the advance of the human being as a toolmaker. The polarity that arose with the proliferation of advanced weaponry in Neolithic times reveals a fundamental split in the human soul itself. We will characterize this split as the tendency to solve life's mysteries (technologist) and the desire to to reveal life's mysteries. Those who wish to solve the mysteries tend toward power as a solution. Those who wish to reveal the mysteries tend toward wisdom. Cain took the path to personal power, while Abel took the path toward

the acceptance of his god's wisdom. When extreme or exclusive, either of these paths can lead to difficulties.

When excessive, the path of power leads to warfare, pollution, and aggressive use of the Earth's resources for personal benefit. When balanced, the path of power can lead to the advancement of knowledge that benefits all humanity. When excessive, the "wisdom of my god" path leads to fundamentalist beliefs and pogroms. When balanced, the "wisdom of my god" path can provide workable cosmological models for the moral tasks of humanity. The obvious task is somehow to create conditions in which to combine the forces of both paths. This requires an approach to science other than reductionism or abstraction, and an approach to religious experience other than dogmatism. Neither approach, when experienced exclusively, includes a cosmology that allows science and religion to unite in a moral perspective of the human role in nature.

Without a progressive cosmology that sees cosmic purpose for the relationship between human beings and the natural world, the possibility of personal transcendent experience, the moral dilemma at the root of the climate crisis, becomes problematic. It will be difficult to resolve the climate crisis with tools that do not allow for a cosmological perspective of the significance of human beings in relation to the life of the Earth. Continuing to view the issues in terms of data available to climate scientists, industry researchers, and special interest groups will probably only perpetuate the mood of impasse that currently dominates the debate. Without a significant cosmology that links human destiny to that of planet Earth and her biography, we are left with only a statistical and computational reference to physical data. No matter how compelling such data may be, it has not inspired much change in the business-as-usual mind-set of contemporary politicians, their patrons, and their constituents.

Much of the split between the human being as part of nature and the human being as simply an observer of nature has to do with the way numbers are used to study and quantify nature. It is possible that the scientific revolution started by the likes of Copernicus, Galileo, Kepler, and Newton has evolved into a condition of abstract cause-and-effect reasoning that prevents humankind from relating morally, and thus effectively, to the climate crisis. In a contemporary cosmos of immense spaces, with Earth as just another mediocre object, has contemporary science sown the seeds of ecological inertia in human beings? Are the present crises of climate, politics, environment, and society, which are deepening in an alarming way, simply mistakes, or is there something deeper behind them for us to consider?

Even in the most ancient times, when science, art, and religion were much closer in purpose, there was a fundamental polarity in the way human beings interacted with the great mystery of the relationship between the humankind and Earth's destiny. In the ancient world, astronomy, architecture, and the formation of calendars were the marks of how the people lived on Earth as a culture—that is, with consciousness that can go beyond mere hunting and gathering. The Egyptians were successful in solving their problems, using geometry to erect buildings and employing phenomenological astronomy to produce calendars. The Egyptians, however, had only rudimentary arithmetic. By contrast, the Babylonians developed place-value arithmetic to a high order, which resulted in accurate astronomical tables for predicting eclipses, making complex calendars, and establishing counting systems, but their knowledge of geometry was undeveloped.

The differences between these two civilizations reveal the possible fundamental polarities in the contemporary process of mathematical scientific inquiry: geometry and arithmetic. In some ancient societies, geometry was the preferred mathematical model for thinking about the world. Geometry requires a more symbolic and metaphysical thinking process. Arithmetic, the foundation of all modern computation, is practical and can be used to create abstractions and cause-and-effect systems of thought. This polarity between metaphysical geometrical reasoning and computational abstract reasoning reflects the worldview of the cultures in which they existed. We can see this polarity even within a single culture. For instance, in Greek times the earlier Platonic worldview of archetypes and a geometrizing universe gave way to the more pragmatic and logical Aristotelian view of numerical categories and abstract reasoning.

Sometimes the two streams come together and a kind of hybrid vigor is the result. Such is the case pictured opposite. The device is known to scholars as the "Antikythera mechanism." It was discovered in the remains of a Roman ship and has been dated to 100 B.C.E. It contained thirty-seven separate gears, which allowed the sailor to compute the date, including leap years, as well as the position of the moon and planets against the zodiac. It could also predict eclipses and the "Saros period" for families of eclipses, which predicts that a similarly positioned eclipse will take place after a cycle of 223 lunar months. The device also carries out subtractions, multiplications, and divisions.

In one elegant mechanism, star groups of the zodiac and the planets moving against them in geometrical, proportional relationships are linked to the abstract demands of computation. Here, the polarities of the more Platonic view—that the planets are the homes of the gods—are linked to the more empirical Aristotelian view—that planets are calculable entities. These poles unite in this celestial computer, built before the time of Christ. The geometry of the ratios and proportions of the mechanism duplicate phenomena seen by the naked eye—the motions of the planets in front of the fixed stars. The computational capabilities of the device supported the more abstract calendrical and arithmetical functions.

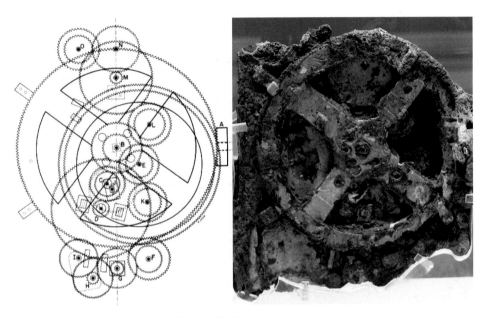

The Antikythera mechanism

The incredible wedding of practicality, knowledge, and technical ability found in the Antikythera device became the focus of the next wave of culture, when Roman engineers and architects made it possible for the Roman war machine to flourish. Geometry, along with the symbolic reverence of metaphysics, was essentially lost to the Romans in favor of computation.

The Arabs of the post-Roman era brought Hindu and Greek scholars to Arabia, where they established great centers of science, literature, religion, and philosophy. Especially in the realm of science, we see the most characteristic energies of the period. A tremendous outpouring of science formalized algebra, algorithms, negative numbers, arithmetic progressions, and geometries that prefigured Fibonacci, and the use of geometric progressions arose in a relatively short period.

However, this flowering began a definite separation of ancient mathematics and mathematics of the new world, especially in astronomy. The computational inheritance that inspired Arabian schools favored abstraction and arithmetical computation as the primary method of science. Only geometers considered geometry useful. Medieval Europe continued the split between the

Johannes Kepler and Isaac Newton

more metaphysical and geometrical Platonists and the rational, categorizing Aristotelians, right up into the Renaissance and beyond.

The contrast between the geometrical–metaphysical pole of Kepler and the emergence of the computational–mathematical brilliance of Isaac Newton perfectly illustrates this ongoing split. Kepler was a highly skilled mathematician who could calculate in his head orbital periods in degrees, minutes, and arc seconds. However, the bulk of his output is in a dense and intricate geometrical reasoning style. His work was deeply geometrical and driven by a search for harmony of the spheres as a kind of religious fervor. The Kepler boyhood home in the town square of a small village outside of Stuttgart stands literally in the shadow of the church that dominates the square. The intricate geometric reasoning, coupled with the sacredness of mystical experience of the harmony of the spheres, brought by Kepler from the ancient world, had to meet the precise phenomenal observations of Tycho Brahe. Without these pragmatic observations, Kepler would have been unable to make his breakthrough. The qualities of religious feeling and profound geometrical capacities that would have endeared Kepler to the ancient world are the exact qualities that have caused some contemporary scholars to diminish his contribution to contemporary science. Some even claim that Kepler found the three laws of the orbits of planets in spite of himself. Scholars have likened his research to a kind of sleepwalking.

It would be tempting to think of Newton as the epitome of the other pole—the rational, calculating, hard-nosed, pragmatic genius. Although it is true that Newton's work marked a threshold from medieval geometric,

symbolic reasoning into the mechanistic and technological present, he himself
was not so convinced that this was the best fit for his insights.

> I likewise call attractions and impulses, in the same sense, Accelerative
> and Motive; and use the words *attraction, impulse,* or *propensity* of any
> sort towards a centre, promiscuously and indifferently, one for another;
> considering those forces not physically, but mathematically: wherefore,
> the reader is not to imagine that by these words I anywhere take upon
> me to define the kind or the manner of any action, the causes or the
> physical reason thereof, or that I attribute forces in a true and physical
> sense, to certain centres (which are only mathematical points); when at
> any time I happen to speak of centres as attracting, or as endued with
> attractive powers. (*Principia,* Definition VIII)

In his own words, the champion of the cause-and-effect physicality that drives
technological physics research denies that the mathematical reasoning he is
discovering has a relation to the physics it spawned. The renowned Newton
scholar Professor Edwin A. Burtt describes this dilemma that the giants of the
scientific revolution faced:

> To get ahead confidently with their revolutionary achievements, they
> had to attribute absolute reality and independence to those entities in
> terms of which they were attempting to reduce the world. This once
> done, all the other features of their cosmology followed as naturally as
> you please. It has, no doubt, been worth the *metaphysical barbarism*
> (author's italics) of a few centuries to possess modern science. Why did
> none of them see the tremendous difficulties involved? Here, too, in the
> light of our study, can there be any doubt of the central reason? These
> founders of the philosophy of science were absorbed in the mathemati-
> cal study of nature.* (emphasis added)

Why would this in itself be the source of "metaphysical barbarism"? Has
not the study of nature from a mathematical view been present since the very
beginning of history? We need to be particularly careful here. Since the transformation of Hindu and Greek mathematics and astronomy into Arabia, a new
impulse had entered the use of mathematics. That impulse was the tendency to
link earthly phenomena (such as the path and velocity of a falling body to the

---

*   Burtt, *The Metaphysical Foundations of Modern Physical Science: A Historical
and Critical Essay* (London: Routeledge, 1924), p. 303.

path and velocity of a celestial body) through an extrapolation based on mathematical calculation. Of course, this is what we just heard Newton deny he was doing. Nevertheless, this is exactly what posterity has taken up from his work as his signature contribution. This says that, everywhere, the prime mover is just the forces and masses of mathematically arranged elements.

This categorizes our Earth as a numerical, quantifiable rock, permeated by forces completely beyond the control of, and sometimes at direct odds with, human soul activity. This exists in a whole other universe from the experience that the Earth is our mother, linked to us through the common consciousness of shared forces. It opens the door to the experience that the cosmos is an ever-expanding, alien dimension of terrible forces ruled by intractable mathematical laws. It also instigates the human will to power to solve the mysteries of those terrible forces and to control and subdue the Earth's forces before they control us. In this shift of consciousness, the mother Earth has somehow become the enemy. How could this happen?

A fundamental feature of the scientific revolution of the late 1700s was the concept that the forces on Earth were similar to those that drive the universe. This led to a search for laws to explain those relationships. As science has progressed, it has also become increasingly difficult to find universal relationships between physical forces and the forces that rule the celestial spheres. This comes to a head in the inability for quantum mechanics to include gravity in its theoretical embrace. The most fundamental celestial element is the very element that ever-finer descriptions of physical forces cannot explain. This is a particularly revealing problem.

Prior to the widespread use of reductionist mathematical techniques, it was understood implicitly that it is possible to identify locations on Earth that have particular resonances to celestial positions. The placement of observatories at particular places on the Earth was a mysterious feature of many ancient cultures.

Why would the ancient peoples spend so much time and effort to create observatories to predict the rising and setting of planets and the motions of the fixed stars? To some researchers, not only were precisely positioned observatories set up to look out to the heavens, but ancients also determined the placement of temples, shrines, and holy places through even more mysterious techniques for finding resonance between celestial coordinates and terrestrial positions.

Stonehenge, in Wiltshire County, U.K.

Those activities, known as *geomancy,* were mystically inspired attempts to link Earth to the movements of planets and stars. Through time, there have been many versions of such techniques. However, except for celestial mechanics and astronomy, none formed a basis for conventional scientific inquiry. Through the development of celestial mechanics, the ability to measure the exact position of a star or planet is quite within the realm of computation.

Trying to project such a position onto the spinning Earth and being able to say with confidence that Jupiter will be over the central Pacific for most of 2006 runs counter to contemporary computational, rational thought. Yet, supported by numerous case studies, this is exactly what the present book describes. How can we track the relationship between planetary movements and an El Niño in a mathematically rigorous way? To do so requires two things. First, a rational, mathematical, and reproducible system of projection must be used so that experiments can be set up and reproduced. Second, some sort of empirical, phenomenological, data-rich information source must be readily available to track and record the results of the mathematical experiments.

The computer and the Internet offer researchers great opportunities for gathering phenomenological climate information. However, setting up a

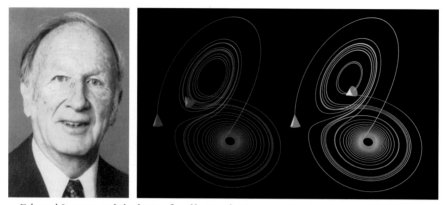

Edward Lorenz and the butterfly effect in the Lorenz attractor: two segments of the three-dimensional evolution of two trajectories (Attraktor, Hamburg)

computational model for climate study is a uniquely difficult task. Edward Lorenz (1917–2008), the father of chaos theory, originally found his insights by trying to model weather systems. He eventually concluded that, in certain ways, climate patterns defy mathematical modeling, even with enhanced computer power. Lorenz worked on slower analog computers and found that the slower more "organic" processing of these early machines could come closer to modeling the intangible interfaces that climate study requires. Even with the slower computational rhythms, Lorenz eventually had to concede that even the tiniest error in the formation of the modeling algorithm grew exponentially through the process of iterating it in the computer many hundreds of times. At some point, chaos ensued as the system sought a higher level of integration. The chaos made it very difficult to model climate systems in flux.

The image above shows the famous Lorenz butterfly, which depicts the polar states that all processes go through to integrate the energies of a whole system. One side of the butterfly represents the positive levels of energy, where the system is functioning and active; then the mid-point between the two "wings" represents the chaos point, where the system loses integrity and begins to unfold on the other side of the butterfly in the negative side of the system's energy economy. This model is used to visualize everything from storms and hurricane development to population explosions and collapses.

There is, however, a division of mathematics that offers modeling potentials for tracking the almost organic, or morphological, transformations of weather systems and climate patterns. That discipline, known as "projective geometry," was invented by Blaise Pascal (1623–1662). Projective geometry has led to some of the most recent advances in present-day geometry, but at the time of its discovery, projective geometry was an outgrowth of the perspective studies of Renaissance artists, coupled to new investigations into optical systems. To make an enormous subject pain-

Blaise Pascal

fully simple, projective systems need a projector. The simplest projector is a point. The point is not just located in space but actively interacts with surrounding fields of activity. When thinking projectively, it helps to use images rather than numbers. The following image shows an optical system interacting with a human eye. Originally, projection was developed with the rise of optics as a geometric discipline.

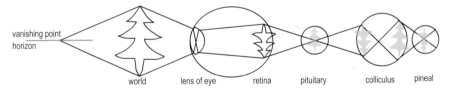

In this optical system, there are several projectors. The vanishing point at the horizon is one. Between the projector and the screen of the world there is an expanding perspective of light rays that is the field of the projector of the vanishing point. This thinking comes from the work of the artist who first discovered the optical laws of perspective in the Renaissance. That research set the stage for the later development of the higher levels of projective space, which became the seed for Bernhard Riemann (1826–1866) and Albert Einstein (1879–1955). The world, or in this instance the tree, has countless points of light on it that reflect in all directions. In the center of the lens of the eye is another point where light rays are gathered. Here, the direction in which the light rays are moving is reversed. The top of the tree is projected down, and the bottom of the tree is projected up to the new screen, the retina.

The image, now upside-down, is gathered again and moved back toward the brain by the action of the nerves. There, another point-like projector, the ganglia of nerves in the area of the hypothalamus/pituitary gland, receives the impulses and, again, up becomes down and vice versa, as the visual nerves split and merge in new combinations as the nerve impulses flow back toward the visual center in the back of the brain.

The condensation of the image into a point and the polar reversal of the image on the expanded field, or screen, is the most fundamental operation in projection. In the image of the Lorenz butterfly, the two wings are the opposite fields or phases of activity of the system studied, whereas the center in the point of chaos, where they reverse, is the attractor, or projector, depending on which side you are on. This is true for sand flowing in an hourglass or for the activity around a black hole in astrophysics. The power of projective geometry to model such disparate systems is a sign of its effectiveness for modeling climate.[*]

The true challenge in climate study is to work with a modeling technique that intrudes least into the chaotic realm of the attractor, or projector. There are two possible approaches to forming such a model. The first is to find an algorithm that, when iterated in a computer, can replicate the observed phenomenon. Lorenz used this approach when he discovered the phase butterfly with the attractor in the center. His conclusion was that the slightest bit of error in the formation of the algorithm leads to great distortion when the algorithm is iterated. This is the essential difficulty in a computational approach to climate study. Lorenz's conclusion was that modeling in this way is next to impossible, owing to the sheer amplitude of variables that could modify the algorithm.

The second approach to climate research is to allow the phenomena themselves to form the iterative patterns and try to find a filter that allows the significant aspects of the system to emerge as the system goes through the complete set of parameters on both sides of the phase butterfly. The problem here, however, is that the chaos of the manifest phenomenon (climate) is too complex for computational mathematics to model it. A possible solution is

---

[*]   For more on projective geometry, see Nick Thomas, *Space and Counterspace: A New Science of Gravity, Time, and Light* (Edinbourgh: Floris Books, 2009). Visit his website at www.nct.anth.org.uk for in-depth and accessible information on this subject.

to use as modeling algorithm mathematical tables (log period tables) that are already predictable, and then reference the natural phenomenon to the periods and fluctuations of the existing table and see if there is a statistical relationship. The most elegant log period tables available are the periods and harmonic phase relationships in an ephemeris, or table of planetary motions. The attempts by Copernicus, Galileo, Kepler, and Newton to understand these most fundamental time phenomena led to the scientific revolution that we celebrate today. The drive for technological advancement has replaced wonder at the incredible richness and precision of the motion-in-arc events of the celestial spheres, which was the very focus of those founders of today's physical science. Perhaps we have thrown out the baby with the bathwater. Or, more likely, we kept the bathwater and forgot that there ever was a baby.

The eclipse grid model is designed to form a phenomenology that allows the variables in the system to form the algorithms that drive the model. To study climate patterns in a context of using projective geometric concepts requires finding a point that can serve as a projector and somehow provide a coincident link between the rhythms of planetary motion-in-arc events and the unfolding of particular weather or climate patterns. The two most fundamental planets to the development of weather events on Earth are the Sun and Moon. The single most important point in the relationship between the Sun and Moon is an eclipse point. By using an eclipse point for the projector, it is possible to see that, on the day of either a solar or a lunar eclipse, significant geometric patterns frequently arise in the placement of lows and highs in the Northern Hemisphere.

The creation of the eclipse grid through geodetic projection allows the construction of a filter that can be used to great advantage when observing weather and climate systems in flux. The eclipse points that form the projector for the grid constantly interact with other planets. Sometimes a planet is approaching an eclipse point. When that happens, a kind of Doppler effect occurs, resulting in a predictable oscillation of high- and low-pressure values on the jet curves associated with the particular eclipse point. The information about the positions of the planets is from an ephemeris, and detailed charts are constructed for every motion-in-arc event that involves an eclipse point.

One constructs a spreadsheet that allows the tracking of each motion-in-arc event and the resulting weather patterns. Over time, it has been possible

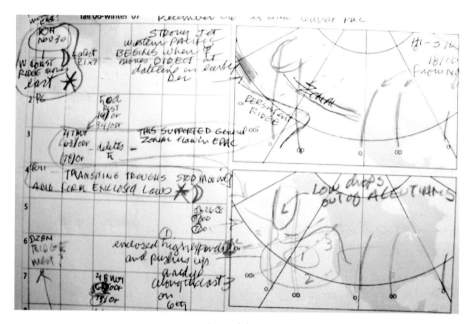

Spreadsheet

to link particular climate patterns to particular combinations of movements (docweather.com/4/show/167/). This data is used to build models and form analogs of historical sequences for climate study. This procedure amounts to a phenomenology of time, allowing the researcher to see into the real-time interaction of planetary-motion sequences and historic-climate responses. The advantage of this approach is twofold. First, no algorithm has to be constructed based on physical data such as dew point, lapse rate, or even SST curves. One neutralizes the disadvantage to driving models with algorithms that are susceptible to errors in the initial inputs. The algorithms used to drive this model are already known. They are the algorithms that are present in the orbital periods of the planets themselves. They are elegant, robust, and, best of all, predictable far into the past for the purpose of study, and far into the future for the purpose of constructing a long-range forecast.

The second advantage is that it is truly remarkable to see the real-time coincidental shifts between motion-in-arc events and weather patterns. The observation of a planet moving in arc on a particular day and a dramatic increase in high pressure or in digging a storm trough is awe-inspiring. This sense of the link between motion-in-arc events of the planets, or the geometry

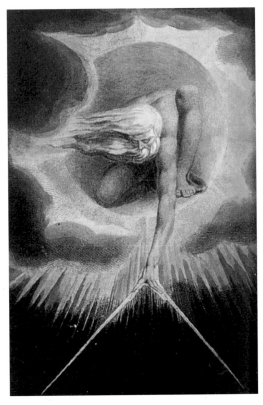

William Blake: God geometrizing

of the eclipse grid and observable phenomena in the weather in exact periods, is the doorway to the cosmological experience that we are not alone on just another rocky planet in an immense soulless void. We live on a planet that is alive and whose soul is composed of the tremendously vital sequences of climate patterns that unite all lands and all humans in one organism. From this cosmological perspective, it may be possible to imagine morally responsible scientific approaches to problem solving where human needs and the needs of the Earth as a living being interact in mutually harmonic ways. From this cosmologically significant point of view, we may, one hopes, be able to say in the future that climate is indeed the soul of the Earth.

GLOSSARY OF TERMS

Angular aspect: the angle that a planet forms with another planet as it travels along the celestial equator in its orbit. The angles are reckoned as parts of a circle.

Bermuda high: a high-pressure air mass normally situated in the western Atlantic. It acts to block the easy passage of storms off of the American continent when it is strong. It can extend into the Gulf of Mexico when it is far to the west. In this position it is the source of the Midwest monsoon

Block: an air mass that has high pressure and is stable on the Earth often blocking the movement of storms. This is also known as a *ridge*.

Decadal influence: This is a climatic pattern that manifests in ten-year intervals.

Degrees of Arc: the number of degrees out of 360 that a planet stands in relation to another planet. A planet at 10° of arc is positioned ten degrees from another planet or from an eclipse point.

Direct motion: a planet moving west to east in celestial longitude is moving direct.

Eclipse line: a line generated from an eclipse point across the North Pole to the opposite longitude. It is the most fundamental line in the planetary flux model.

Eclipse points: a point in celestial longitude that is the place where the Sun is situated on the day of either a solar or lunar eclipse. Eclipse points are the key feature for setting up eclipse grids in the planetary flux model.

Geodetic projection or equivalency: is a technique that enables a position in the heavens to be projected down onto the Earth at a specific terrestrial latitude and longitude.

High pressure: a dense mass of air that tends to remain in place as warmer air circulates around it. High-pressure areas are also called *ridges*.

Jet curve: a curved harmonic area found by drawing a 45° or 72° circle using an eclipse point as the center. It is a preferred place for the jet stream to change direction.

Jet stream: a high-altitude river of rapidly moving air that steers storms.

Low pressure: a mass of air moving upward due to its warmth. The wind circulation around a low is counterclockwise in the Northern Hemisphere. Lows bring storm energies with them. They are also called *troughs*.

Meridional flow: the movement of the transcontinental jet stream vertically in a north-to-south pattern following the longitudinal meridians.

Motion-in-arc: the motion of a planet in celestial longitude. Each new degree of motion into a new longitude consists of a motion-in-arc. Motion-in-arc is the primary driving force for the planetary flux model. The motion-in-arc of a planet is often more important in this system than its actual position,

Nodal cycle: the motion of the lunar node backward through the zodiac over an 18.6-year rhythm.

Node: where the path of the Moon crosses the Sun's ecliptic to form an eclipse.

Planetary flux model: a system of climatology that uses the motion-in-arc of planets moving across a grid of lines projected from an eclipse position as the basis for predicting weather changes.

Polar 90: a zone of disturbance that is projected from points that stand at angles of arc of 90° to each eclipse point. Each 90° point projects a line to the North Pole, and the area between these lines is often a determining zone in the climate patterns of a given period.

Reflex point: a position 180° from the actual eclipse position. In the planetary flux model, reflex points have similar qualities to eclipse points, even though there is not actual planetary presence in the celestial hemisphere.

Retrograde motion: a planet moving from east to west in celestial longitude is moving retrograde, or looping.

Ridge: a high-pressure air mass that blocks the motion of the jet stream.

Sidereal astrology: the practice of using the stars as the reference point for reckoning celestial longitude rather than employing the tropical system, which uses the placement of the Sun at the solstice.

Squeeze: A squeeze arises between one planet moving direct and another planet moving retrograde into each other in close proximity. Sea surface temperatures (SST) rise in the area between the two planets.

Tandem transit: this occurs when two planets approach or cross a point within a few degrees of each other. It also refers to a situation in which two planets cross the two eclipse points simultaneously.

Teleconnections: a situation in which a large high-pressure area links across a vast space to another high-pressure area. The circulation around each one moves between the two of them to cover large areas.

Transit: this occurs when one planet passes another in longitude or passes a particular point in longitude.

Trough: a low-pressure curve in the jet stream that steers storms around blocks.

Zonal flow: the movement of storms when the storm jet follows the latitudinal lines horizontally across the top of the United States.

INDEX

Articles

Writings presently available include "Earth Consciousness" (an image of stars falling out of the sky as snow and then falling to earth as crystals); "The Alchemical Mandala" (a dynamic map of developmental stages for self-transformation); "The Heart of Clay" (intensive soil cultivation and the addition of proper soil amendments). "Tree Painting as an Esoteric Deed" (dealing with tree-paste procedures for biodynamic gardeners); "The Alchemical Worldview" (deals with the central task of the human being: to achieve what is known as the second birth, or what is known to students of Rudolf Steiner as the birth of the "I" being).

Art Prints

Dennis's art is available as prints in various sizes and as cards.

"Bighorn Mountains" (left); "Coastal Rhythms" (right)

Lecture Downloads

Numerous lectures are available, including "Agricultural Alchemy" (on principles founded by the alchemists that can serve as the basis for a new way of viewing natural forces); "Black Madonna" (tracing the concept of the Madonna from Gnostic sources as the basis for meditative practices); "Creativity in Teaching" (on how to avoid burnout when dealing with the demands of students, colleagues, and parents); "Principles of the Biodynamic Preparations" (workshop on the principles of the Biodynamic preparations given at Frey Vineyards).

The Work of Dennis Klocek as "Doc Weather"
docweather.com

About Doc Weather

DocWeather.com is a unique website based on a climatic technique that interfaces daily weather data from the National Weather Service into a geometrical system that is the result of twenty-four years of experimental work. In DocWeather, precise weather observation and advanced planetary geometry are linked to historic climatic patterns across the U.S.

Doc Weather Basics

Explore further the basis of this book, *Climate: Soul of the Earth.* Over the years, DocWeather's system has revealed the observable correlation between the Earth's climate and the planets, the Sun, and the Moon, including their angles in reference to one another and their location in the sky "over" the Earth. The use of planets and constellations on the site is merely the clearest way to reference these positions and angles. From this starting point, articles on the site discuss the energies those positions and angles represent and how they impact our climate and weather.

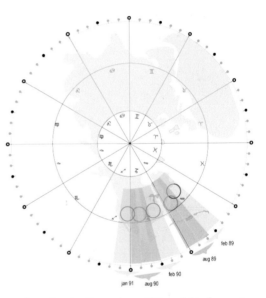

Chart for the discussion of "decadal influence"

The Weather Eye

In-depth articles ranging from eclipse patterns and the music of the spheres, to case studies of record freezes, floods, and droughts. Some of these articles formed the basis for chapters in this book, while others further explore various aspects of weather and climate phenomena.